POLY (VINYL ALCOHOL)[PVA]-BASED POLYMER MEMBRANES

POLY (VINYL ALCOHOL)[PVA]-BASED POLYMER MEMBRANES

SILVIA PATACHIA
ARTUR J. M. VALENTE
ADINA PAPANCEA
AND
VICTOR M. M. LOBO

Nova Science Publishers, Inc.
New York

Copyright © 2009 by Nova Science Publishers, Inc.

All rights reserved. No part of this book may be reproduced, stored in a retrieval system or transmitted in any form or by any means: electronic, electrostatic, magnetic, tape, mechanical photocopying, recording or otherwise without the written permission of the Publisher.

For permission to use material from this book please contact us:
Telephone 631-231-7269; Fax 631-231-8175
Web Site: http://www.novapublishers.com

NOTICE TO THE READER

The Publisher has taken reasonable care in the preparation of this book, but makes no expressed or implied warranty of any kind and assumes no responsibility for any errors or omissions. No liability is assumed for incidental or consequential damages in connection with or arising out of information contained in this book. The Publisher shall not be liable for any special, consequential, or exemplary damages resulting, in whole or in part, from the readers' use of, or reliance upon, this material.

Independent verification should be sought for any data, advice or recommendations contained in this book. In addition, no responsibility is assumed by the publisher for any injury and/or damage to persons or property arising from any methods, products, instructions, ideas or otherwise contained in this publication.

This publication is designed to provide accurate and authoritative information with regard to the subject matter covered herein. It is sold with the clear understanding that the Publisher is not engaged in rendering legal or any other professional services. If legal or any other expert assistance is required, the services of a competent person should be sought. FROM A DECLARATION OF PARTICIPANTS JOINTLY ADOPTED BY A COMMITTEE OF THE AMERICAN BAR ASSOCIATION AND A COMMITTEE OF PUBLISHERS.

LIBRARY OF CONGRESS CATALOGING-IN-PUBLICATION DATA
Available upon request

ISBN: 978-1-60692-384-9

Published by Nova Science Publishers, Inc. ✦ New York

Contents

Chapter 1	Introduction	1
Chapter 2	Separations by Membranar Processes	7
Chapter 3	Other Domains of Membranes Application	43
Chapter 4	Conclusion	75
References		77
Index		89

Chapter 1

INTRODUCTION

Poly(vinyl alcohol) (PVA) is a polymer of great interest because of its many desirable characteristics specifically for various pharmaceutical, biomedical, and separation applications. PVA has a relatively simple chemical structure with a pendant hydroxyl group (figure 1a). The monomer, vinyl alcohol, does not exist in a stable form, rearranging to its tautomer, acetaldehyde. Therefore, PVA is produced by the polymerization of vinyl acetate to poly(vinyl acetate) (PVAc) followed by the hydrolysis to PVA (figure 2). Once the hydrolysis reaction is not complete, there are PVA with different degrees of hydrolysis (figure 1b). For practical purposes, PVA is always a co-polymer of vinyl alcohol and vinyl acetate [1].

PVA must be cross-linked in order to be useful for a wide variety of applications. A hydrogel can be described as a hydrophilic, cross-linked polymer, which can sorbe a great amount of water by swelling, without being soluble in water. Other specific features of hydrogels are their soft elastic properties, and their good mechanical stability, independent of the shape (rods, membranes, microspheres, etc.).

PVA can be prepared by chemical or physical cross-linking; general methods for chemical cross-linking are the use of chemical cross-linkers or the use of electron beams or γ-radiation, whilst the most common method to produce physical cross-linking PVA is the so-called "freezing-thawing" process.

PVA can be cross-linked [2] using cross-linking agents such as glutaraldehyde, acetaldehyde, etc. When these cross-linking processes are used in the presence of sulfuric acid, acetic acid, or methanol, acetal bridges form between the pendant hydroxyl groups of the PVA chains. As with any cross-linking compound, however, residual amounts are present in the PVA gel matrix;

furthermore, other compounds such as initiators and stabilisers will reamin after synthesis. To use these gels for pharmaceutical or biomedical applications, we will have to extract all residues from the gel matrix. This is an extremely undesirable time-consuming extraction process; also, if the process is not 100 % efficient and the residue is not completely removed, the gel will not be acceptable for biomedical or pharmaceutical applications.

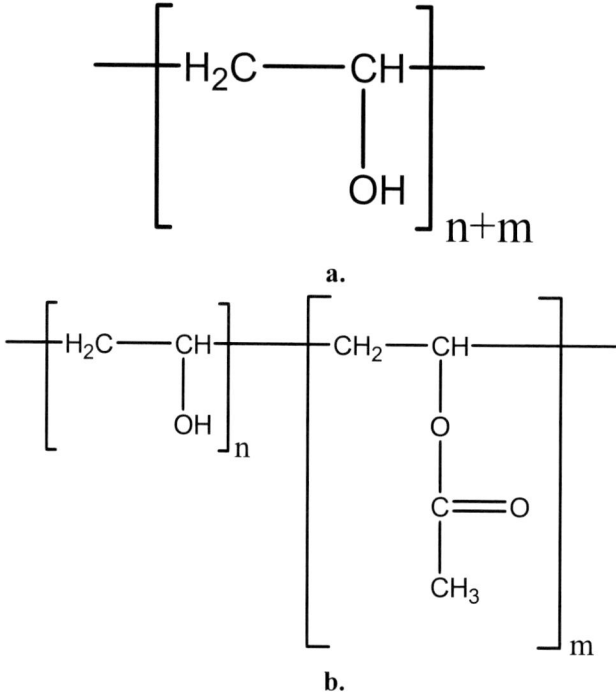

Figure 1. Molecular strucuture of PVA fully hydrolyzed (a) and partially hydrolyzed (b).

Other methods of chemical cross-linking include the use of electron beam or γ-irradiation. These methods have advantages over the use of chemical cross-linking agents as they do not leave behind toxic, elutable compounds. The minimum gelation dose of γ-rays (from ^{60}Co sources) depends on the degree of polymerisation and the concentration of polymer in solution [3]. The effect of irradiation dose on the physical properties of PVA fibers, hydrogels and films irradiated in water is reported in [4-6].

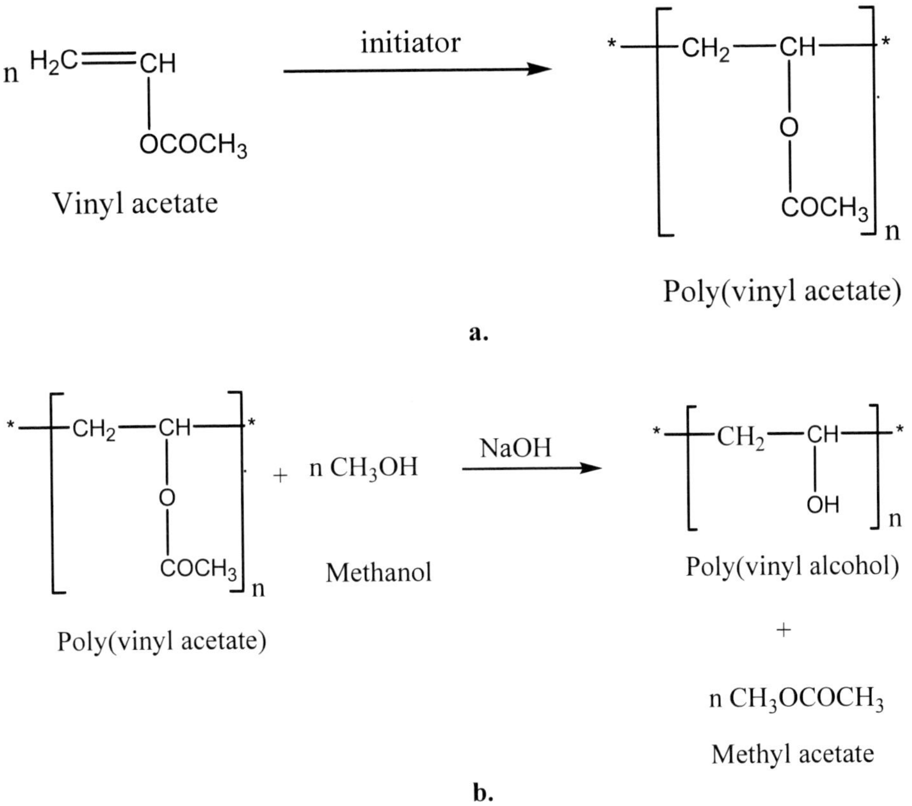

Figure 2. Polymerization of vinyl acetate (a) and hydrolysis of PVAc to PVA (b).

The third mechanism of hydrogel preparation involves "physical" crosslinking due to crystallite formation [7]. This method addresses toxicity issues because it does not require the presence of a cross-linking agent (figure 3). Such physically cross-linked gels also exhibit higher mechanical strength than PVA gels crosslinked by chemical or irradiative techniques because the mechanical load can be distributed along the crystallites of the three-dimensional structure [1]. Some characteristics of these "physically" crosslinked PVA gels include a high degree of swelling in water, a rubbery and elastic nature, and high mechanical strength. In addition, the properties of the gel may depend on the molecular weight of the polymer, the concentration of the aqueous PVA solution, the temperature and time of freezing and thawing, and the number of freezing/thawing cycles [8-10].

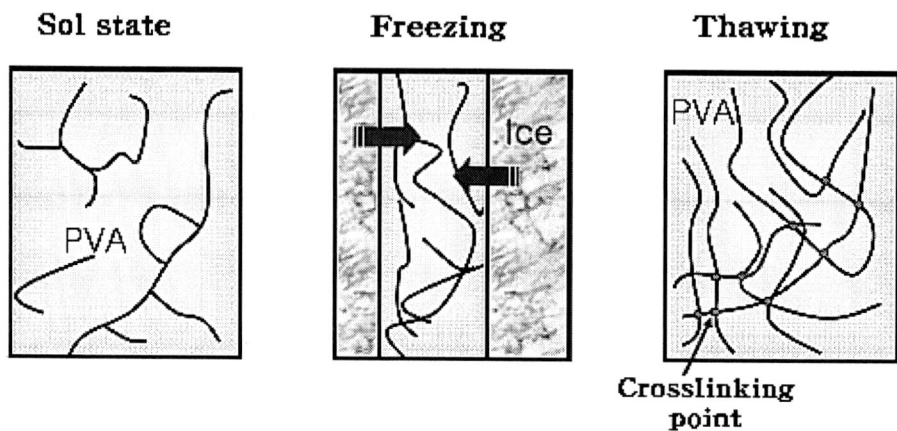

Figure 3. Schematic representation of PVA gels formed by freezing/thawing process.

PVA hydrogels have been used for numerous biomedical and pharmaceutical applications. PVA hydrogels are non-toxic, non-carcinogenic, show bioadhesive characteristics, and they are easily processed. The safety of PVA is based on the fact that the acute oral toxicity of PVA is very low, with LD50s (the amount of a material, given all at once, which causes the death – lethal dose - of 50 % of a group of test animals) in the range of 15-20 g/kg; when orally administered PVA is very poorly absorbed from the gastrointestinal tract; PVA does not accumulate in the body when administered orally; PVA is not mutagenic or clastogenic; and the no-observed adverse-effect level (NOAEL) of orally administered PVA in male and female rats were 5000 mg/kg body weight/day in the 90-day dietary study and 5000 mg/kg body weight/day in the two-generation reproduction study, which was the highest dose tested [11].

Furthermore, PVA gels exhibit a high degree of swelling in water and a rubbery and elastic nature. For all these features PVA is an excellent biomaterial. In fact, PVA is capable of simulating natural tissue and can be readily accepted into the body. PVA gels have been used for contact lenses, the lining for artificial organs, and drug-delivery applications. Recently, intelligent hydrogels have been used to produce micro- and nano-fabricated devices that seek to develop a platform of well controlled functions in the micro- and nano-level. For example, polymer surfaces in contact with biological fluids, cells, or cellular components can be tailored to provide specific recognition properties or to resist binding depending on the intended applications. Another recent application of PVA is related with the development of biomimetic methods to build biohybrid systems or even biomimetic materials for drug delivery, drug targeting, and tissue engineering devices. Besides all these applications, PVA is an important gel in different enginnering and industrial fields. For example, in the U.S.A., the majority of PVA is used in the textile industry as a sizing and finishing agent. PVA can also be incorporated into a water-soluble fabric in the manufacture of degradable protective apparel, laundry bags for hospitals rags, sponges, sheets, covers, as well as physiological hygiene products. PVA is also widely used in the manufacture of paper products. As with textiles, PVA is applied as a sizing and coating agent. It provides stiffness to these products making it useful in tube winding, carton sealing and board lamination. PVA is used as a thickening agent for latex paint and common house hold white glue or in other adhesive mixtures such as remoistenable labels and seals, as well as gypsum-based cements such that used for ceramic tiles. PVA is relatively insoluble in organic solvents and its solubility in aqueous solutions is adaptable to its necessary application [11].

The US Food and Drug Administration (FDA) allows PVA for use as an indirect food additive in products which are in contact with food [11]. For example, under 21 CFR 73.1, PVA is approved as a diluent in color additive mixtures for coloring shell eggs and under 21 CFR 349.12, PVA is approved as an ophthalmic demulcent at 0.1–4.0 %.

Other applications of PVA are in areas of water and wastewater treatment (extraction, ultra-filtration, ion-exchange materials, etc.), catalysis, separation, etc.

As an industrial and commercial product, PVA is valued for its solubility and biodegradability, which contributes to its very low environmental impact. Several microorganisms ubiquitous in artificial and natural environments — such as septic systems, landfills, compost and soil — have been identified and they are able to degrade PVA through enzymatic processes.

Membranes have gained an important place in chemical technology and are used in a broad range of applications. The key property that is exploited is the ability of a membrane to control the permeation rate of a chemical species through the membrane. In controlled drug delivery, the goal is to moderate the permeation rate of a drug from a reservoir to the body. In separation applications, the goal is to allow one component of a mixture to permeate the membrane freely, while hindering permeation of other components.

The objective of this chapter is to give an overview of the developments in synthesis and applications of PVA-based membranes in the last years.

Chapter 2

SEPARATIONS BY MEMBRANAR PROCESSES

2.1. PERVAPORATION PROCESSES

Pervaporation, in its simplest form, is an energy efficient combination of membrane permeation and evaporation. Pervaporation involves the separation of two or more components across a membrane by differing rates of diffusion through a thin polymer and an evaporative phase change comparable to a simple flash step. A concentrate and vapour pressure gradient is used to allow one component to preferentially permeate across the membrane. A vacuum applied to the permeate side is coupled with the immediate condensation of the permeated vapors. Pervaporation is typically suited to separating a minor component of a liquid mixture, thus high selectivity through the membrane is essential.

Despite concentrated efforts to innovate polymer type and tailor polymer structure to improve separation properties, current polymeric membrane materials commonly suffer from the inherent drawback of tradeoff effect between permeability and selectivity, which means that membranes more permeable are generally less selective and vice versa.

Pervaporation (PV) is considered to be a promising alternative to conventional energy intensive technologies like extractive or azeotropic distillation in liquid mixtures' separation for being economical, safe and ecofriendly. PV can be considered the so-called 'clean technology', especially well-suited for the treatment of volatile organic compounds. The separation of compounds using pervaporation methods can be classified into three major fields viz. (i) dehydration of aqueous–organic mixtures [12], (ii) removal of trace volatile organic compounds from aqueous solution [13] and (iii) separation of organic–organic solvent mixtures [14]. The hydrophilic membranes were the first ones to have found an industrial application for organic solvent dehydration by PV [15]. Very recently, B. Smita et al. [16] reported that some restrictions for a variety of membranes for their application are still encountered, suggesting potential routes to overcome these drawbacks as, for example, the development of appropriate membrane material (flux and selectivity of a membrane are deciding factors in pervaporation mass transport; therefore, development of a new polymer material is a key research area in membrane technology. The aim in the development of new pervaporation membranes is either to increase the flux, keeping the selectivity constant or aiming for higher selectivities at constant flux, or both. In order to achieve such goals, the use of PVA as component of copolymers, blends, or composites membranes for pervaporation has been used.

2.1.1. Pervaporation of Phenol/Water

Using pervaporation through PVA membranes, J. W. Rhim et al. [17] have studied the separation of water-phenol mixtures.

The pervaporation separation of water-phenol mixtures was carried out using poly(vinyl alcohol) (PVA) cross-linked membranes with low molecular weight poly(acrylic acid) (PAA), at 30, 40, and 50 °C. They have used pervaporation because the separation rate is higher (for liquid organic mixtures) in pervaporation than in reverse osmosis.

A separation factor of the mixture, α, is calculated using

$$\alpha = (Y_{water} / Y_{phenol}) / (X_{water} / X_{phenol})$$

where X is the weight fraction of permeate and Y, the weight fraction of feed.

A very high separation factor has been obtained in phenol dehydration by using pervaporation process and PVA/PAA as membranes. The membrane composition and the process characteristics are presented in table 1.

Conclusion: the separation factor increases by increasing the cross-linker, and decreases by increasing the temperature.

2.1.2. Isopropanol/Water Separation

The selective separation of water from aqueous solutions of isopropanol or the dehydration of isopropanol can be carried out with different membranes, which contain polar groups, either in the backbone or as pendent moieties. For the dehydration of such a mixture, poly(vinyl alcohol) (PVA) and PVA-based membranes have been used extensively. PVA is the primary material from which the commercial membranes are fabricated and has been studied intensively for pervaporation because of its excellent film forming, high hydrophilicity due to – OH groups as pendant moieties, and chemical-resistant properties. On the contrary, PVA has poor stability at higher water concentrations, and hence selectivity decreases remarkably.

Table 1. Characteristics of the separation process by pervaporation function of the membrane composition and structure, composition of feed mixture and temperature [18]

Membrane composition PVA/PAA	Composition of liquid mixture		Permeation rate / / (g m^{-2} h^{-1})	T / °C	Separation factor
80/20	phenol/water	80/20	50	30	3580*

* Ref. 17.

The use of conjugated polymer as membranes to separate various liquid mixtures has been reported in the literature [19,20]. From those, polyaniline (PANi) is one of the most interesting and studied conjugated polymers. Polyaniline is usually prepared by direct oxidative polymerization of aniline in the presence of a chemical oxidant, or by electrochemical polymerization on different electrode materials [21,22]. The possible interconversions between different oxidation states and protonated and depronated states [23], figure 4, make this material remarkable for different purposes. Under most conditions, PANi acts as a passive material, but electrolysis or exposure to acidic aqueous solutions gives rise to conductive materials. In fact, the susceptibility of PANi protonation-deprotonation is an important property once it makes possible to control the electrical conductivity of polyaniline, being possible to obtain changes of more than two orders of magnitude in the electrical conductivity [24]. Both synthesis and characterization of PANi have been reviewed by different authors [21-23]. These reviews deal with chemical, electrochemical and gas-phase preparations, polymerization mechanisms, physicochemical and electrochemical properties, redox mechanisms, theoretical studies, and applications of the polymer.

Interest in polyaniline (PANi), as a material for membrane separations, stems for its high selectivity toward liquids since most liquids are in the size regime of 0.2–1 nm. Another advantage is that PANi has the ability to be tailored after its synthesis through doping/undoping processes. Since there is a tremendous driving force for adding protonic dopants to the imine nitrogens in the PANi backbone [20], the polymer chains are readily pushed apart by the incoming dopants. Thus, doping would induce morphological changes in the polymer resulting in varying permselectivities. Besides such morphological changes, the undoped and doped forms of PANi exhibit different characteristics. For instance, the undoped form of PANI is hydrophobic, while the doped form is hydrophilic [25,26]. Hence, doped PANi preferentially permeates water over the organics, such as isopropanol. The above-mentioned advantages are considered to search for novel membranes containing PANi nanoparticles dispersed in the PVA matrix. The synthesis of a novel hybrid nanocomposite membrane by in situ polymerization of aniline in the PVA matrix in acidic media is described in the Ref. 27. Aniline monomer was introduced into the PVA matrix and by carrying in situ polymerization outside the mesopores of the polymer matrix, a nanocomposite structure was formed. The organic phase extends along the channels to the openings in the nanocomposite structure due to strong interactions between the nanoparticle formed and the continuously polymerized PANi nanoparticles. This hybrid polymer shows lower swelling degree and higher water selectivity (about five-folds) compared to the plain poly(vinyl alcohol).

M. Sairam et al. [28], taking on the basis of the cited PANi nanoparticles dispersed in the PVA matrix, suggests the incorporation of TiO_2 filler-coated with polyaniline (emeraldine state) salt nanoparticles in PVA. PVA contains a large number of hydroxyl groups which can effectively inhibit the aggregation of TiO_2 nanoparticles by the organic surface modification and help to keep the TiO_2 particles well dispersed in the aqueous PVA solution at the nano-scale for dehydration of iso-propanol. In order to control the dispersion of TiO_2 fillers and to adjust the permselectivity, the PV membranes formed have been crosslinked chemically with glutaraldehyde. With this modification of TiO_2 nanoparticles, it is expected that strong interfacial bond, viz., Ti–O–C be formed on the surface of TiO_2 nanoparticle, anchors PVA molecules to the surface of TiO_2 nanoparticles such that surfaces of TiO_2 nanoparticles will be wrapped with the layer of PVA polymer. It is known [29] that there are number of Ti–OH groups that will cover the surface regions of TiO_2 nanoparticles. When PVA chain segments are adsorbed onto the surface of TiO_2 nanoparticle, Ti–OH groups on the surface of TiO_2 nanoparticles will react with the hydroxyl groups linking to the PVA chains. Dehydration and condensation reactions can occur between both the hydroxyl groups.

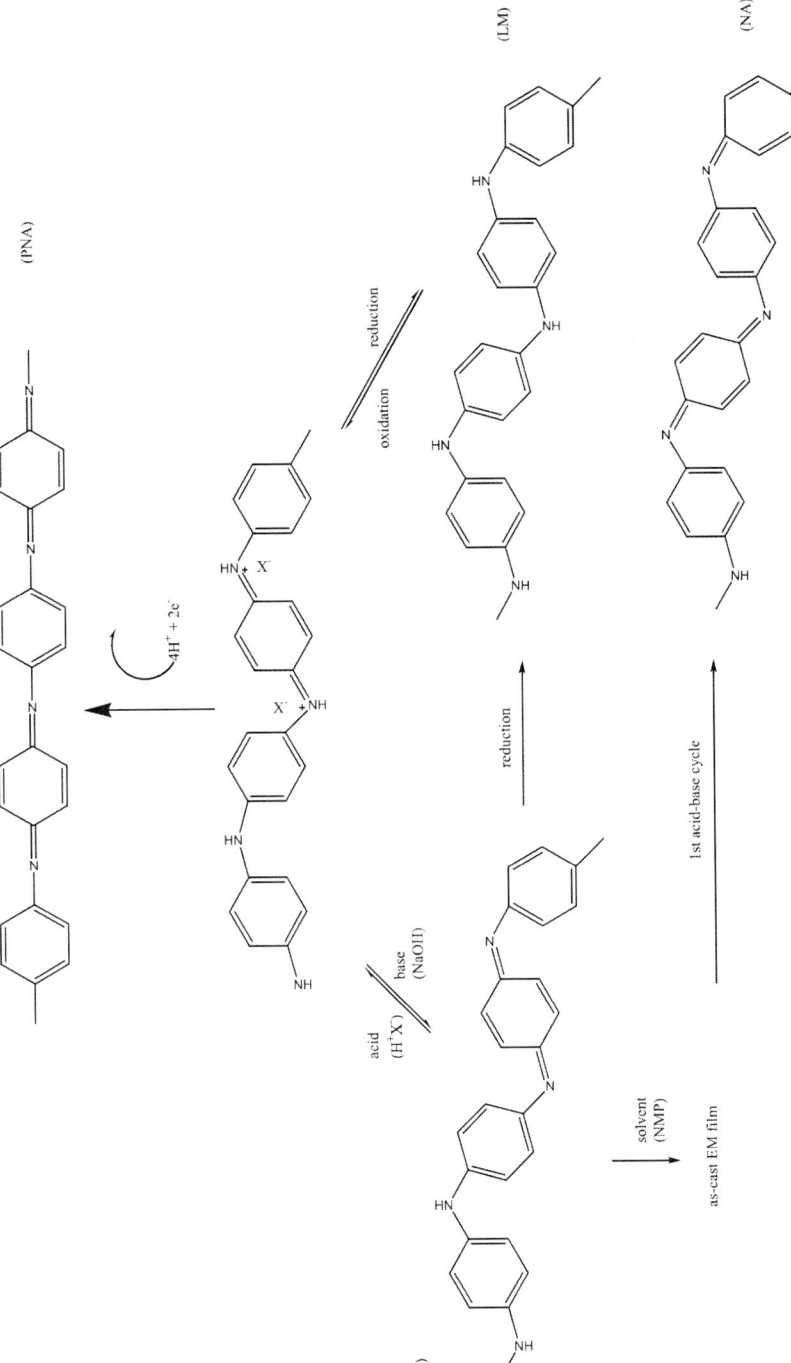

Figure 4. Interconversions among the various intrinsic oxidation states and protonated/deprotonated states in polyaniline.

Another approach to enhance separation performance of membrane for dehydration of isopropanol is the modification of PVA membranes in gaseous plasma [30]. The modification of membrane properties in nitrogen plasma environment lead to increase in selectivity by about 1477 at 25 °C; such increase in the selectivity is justified by an increase of cross-linking on membrane surface provoked by plasma treatment.

The same authors reported the possibilities of using a membrane made by PVA modified by LiCl, whose surface has been modified by exposure to low-pressure nitrogen plasma [31]. The best results have been obtained for 0.05 wt% of LiCl in PVA membrane at 25 °C (selectivity 14 and flux 250 g m^{-2}h^{-1}).

Hybrid membranes composed of poly(vinyl alcohol) (PVA) and tetraethylorthosilicate (TEOS), synthetised via hydrolysis and a co-condensation reaction for the pervaporation separation of water-isopropanol mixtures has also been reported [32]. These hybrid membranes show a significant improvement in the membrane performance for water–isopropanol mixture separation. The separation factor increased drastically upon increasing the crosslinking (TEOS) density due to a reduction of free volume and increased chain stiffness. However, the separation factor decreased drastically when PVA was crosslinked with the highest amount of TEOS (mass ratio of TEOS to PVA is 2:1). The highest separation selectivity is found to be 900 for PVA:TEOS (1.5:1 w/w) at 30°C. For all membranes, the selectivity decreased drastically up to 20 mass % of water in the feed and then remained almost constant beyond 20 mass %, signifying that the separation selectivity is much influenced at lower composition of water in the feed.

Recently, a new effective membrane for different organic solvents dehydration by pervaporation has been reported. Novel hydrophylic polymer membranes based on crosslinked poly(allylamine hydrochloride) (PAA.HCl)-PVA have been developed [33]. The crosslinking agent was GA. The role of the PVA into the membrane is to increase its flexibility and the stability. But the increasing of the PVA ratio, determines the decreasing of the water selective amine hydrochloride functional groups amount and as consequence, the rate of water intake by the membrane decreases. So, for different specific applications the optimization of the PAA.HCl/PVA ratio in the formulation is essential. Also, the amount of GA and curing temperature has to be optimized to obtain the desired membrane properties. The characteristics of the iso-propanol (IPA) dehydration process, by using the pervaporation technique, are presented in the table 2.

Table 2. The characteristics of the iso-propanol (IPA) dehydration process, by using pervaporation technique, through (PAA.HCl)-PVA membrane

Composition of the membrane PAA.HCl/PVA/GA	Composition of liquid mixture		Water flux / / (kgm^{-2}h^{-1})	T / °C	Separation factor
60/35/5	IPA-water (wt%)	85/15	3.14	70	2930

* Ref. 33.

2.1.3. Pervaporation Ethanol/Water

Alcohol is a clean energy source that can be produced by the fermentation of biomass. However, it needs to be highly concentrated. In general, aqueous alcohol solutions are concentrated by distillation, but an azeotrope (96.5 wt% ethanol) prevents further separated by distillation. Pervaporation, a membrane separation technique, can be used for separation of these azeotropes: pervaporation is a promising membrane technique for the separation of organic liquid mixtures such as azeotropic mixtures [34] or close-boiling point mixtures.

The synthesis of novel organic-inorganic hybrid membranes via hybridization between organic and inorganic materials using the sol-gel reaction is reported elsewhere [35]. It is well-known that poly(vinyl alcohol) (PVA) membranes are highly water permselective for aqueous ethanol solutions during PV. However, the swelling of the PVA membrane in an aqueous ethanol solution results in both an increase in solubility and diffusivity of ethanol, and consequently lowers the water permselectivity [36]. The control of membrane swelling has been attempted by cross-linking, surface modification, and annealing methods. However, it is difficult to effectively control the swelling of the membrane. An attempt to improve and to control the swelling is done by using mixtures of PVA and tetraethoxysilane (TEOS), as an inorganic component, in order to obtain PVA/TEOS hybrid membranes prepared by sol-gel reaction. The addition of TEOS into the PVA membrane decreases the swelling of the membrane and improves the water permselectivity of the PVA/TEOS hybrid membrane. T. Uragami *et al.* also studied the effect of the annealing process to PVA/TEOS hybrid membranes. They found that the separation factor H_2O/EtOH increases from 329 to 893 (with the same permeation rate) when PVA/TEOS (TEOS content 25 %) membranes are submitted from an annealing process at 160 °C during 8 hours to 130 °C during 24 hours.

In a previous section, the effect of plasma on PVA surface for pervaporation processes was also mentioned. In fact, plasma treatment is a surface-modification method to control the hydrophilicity–hydrophobicity balance of polymer materials in order to optimize their properties in various domains, such as adhesion, biocompatibility and membrane-separation techniques. Non-porous PVA membranes were prepared by the cast-evaporating method and covered with an allyl alcohol or acrylic acid plasma-polymerized layer; the effect of plasma treatment on the increase of PVA membrane surface hydrophobicity was checked [37]. The allyl alcohol plasma layer was weakly crosslinked, in contrast to the acrylic acid layer. The best results for the dehydration of ethanol were obtained using allyl alcohol treatment. The selectivity of treated membrane (H_2O wt% in the pervaporate in the range 83–92 and a water selectivity, α_{H2O}, of 250 at 25 °C) is higher than that of the non-treated one ($\alpha_{H2O} = 19$) as well as that of the acrylic acid treated membrane ($\alpha_{H2O} = 22$).

PVA dense membranes treated by acrylic acid (Acr.Ac) plasma were obtained by A. Essamri *et al.* [38]. These membranes were used for dehydration of the EtOH-H_2O mixtures by pervaporation. The behaviour of these films on ethanol-water pervaporation has increased performances after plasma treatment.

This means an increase of the flux (J) and water selectivity (β) for the modified membrane – due to the surface properties modification by plasma treatment – comparing to the untreated membrane.

Conclusion: using plasma treatment, a good ratio between flux and selectivity could be obtained.

Also, different other techniques for obtaining PVA/PAcr.Ac blends were reported [39-56]:

1. mixing of PVA and PAcr.Ac. aqueous solutions, solvent evaporation and thermal treatment of the resulted film;
2. mixing of PVA and PAcr.Ac. aqueous solutions with curing agents solutions, solvent casting and thermal treatment of the resulted film;
3. repetitive cycles of freezing and thawing of the aqueous solutions of polymer mixture, in the presence or the absence of the curing initiators;
4. UV irradiation either of the aqueous solutions of polymers mixtures in the presence of photoinitiators and curing agents, or of the PVA hydrogels swelled with acrylic acid;
5. acrylic acid polymerization in the matrix of PVA, in the presence of curing agents and initiators;
6. sedimentation polymerization of acrylic acid in the presence of PVA solution, crosslinking agent and initiator;

7. swelling of PAcr.Ac dried hydrogels with an aqueous solution of PVA and application of the freezing and thawing technique.

Changes in blends' properties were obtained through different ways: by changing the polymer mixing ratios; by using a small amount of acids that are catalysts for esterification reactions; by changing the crosslinking degree of the polymers.

PVA and PAcr.Ac. are compatible polymers on the whole range of composition [40-45].

These blends are homogeneous and the films are transparent evidencing a good clarity [43,57] or semitransparency [43]. However a heterogeneous IPN [45] was obtained by acrylic acid polymerization in a PVA matrix.

The blends could exhibit different morphologies: continuous or microporous [46].

The blend crystalinity degree decreases with increasing the PAcr.Ac content up to 50 wt%, from 26 % to 2 %, and then remains constant [42].

The blends are water insoluble [47]. They can swell in different solvents: water, acetone, aqueous solutions of acids and alkalis [47].

The swelling ratios increase with increasing the PAcr.Ac content in IPNs [45,48].

It was pointed out that the swelling degree evidenced a strongly decrease as the PAcr.Ac content in membrane decreases to 20% [49].

The technique of obtaining blends influences their swelling ratios by inducing different crosslinking degrees. For example, increasing the number of freezing-thawing cycles leads to a swelling ratio significant decrease [50].

In general, the swelling ratios increase with the increasing the temperature up to 40 °C [45]. The dependence of swelling ratios on temperature shows a different function of the blend composition.

So, interpenetrating polymer networks (IPNs) with a weight ratio of vinyl alcohol residue in PVA to acrylic acid monomer 4:6 exhibit positive swelling changes with temperature but IPNs 6:4 evidence negative swelling ones [48].

pH strongly influences the swelling behavior of the blends. For example, the difference of the swelling ratio of IPN 4:6 between pH=4 and pH=7 is 2.0 [48].

Membranes of PVA/PAcr.Ac blends evidence a selective permeability against different components of a liquid mixture. So, they may be used for the ethanol dehydration by pervaporation technique.

Table 3 presents a summary of the published results.

These blends show good mechanical properties. The presence of PVA in the blend improves the mechanical properties. Hydrogels have a significant mechanical strength and elasticity [42]. The tensile strengths are larger than those of crosslinked PVA membranes and show a maximum value at about 0.7 wt% glutaric dialdehyde (GA) [46].

Table 3. Characteristics of the separation process by pervaporation function of the membrane composition and structure, composition of feed mixture and temperature [18,43,46]

Composition of the membrane PVA/PAcr.Ac	Composition of liquid mixture ethanol-water	Permeation rate / (g m^{-2} h^{-1})	T / °C	Separation factor	Permeate activation energy E_a / (kJ mol^{-1})
50/50 IPN	95.6/4.4	260	50	50 12	-
30/70 IPN	10/90 85/15	5000 750	50	0.8 15	30.5
50/50 IPN	10/90 85/15	3800 360	50	0.85 15	30.9
70/30 IPN	10/90 85/15	2700 110	50	1.0 18	-
90/10 IPN	10/90 85/15	2000 90	50	3.0 39	38.9
80/20	95.6/4.4	9 30	60 75	14000 5800	-
	90/10	27 60	60 75	9000 2800	
	80/20	65 125	60 75	1500	
	50/50	120	75	260	
95/5		550	60	150	

Table 4. Dehydration of ethanol, using membranes PAA.HCl (60 wt%)–PVA (35 wt%–GA (5 wt%) (aprox. 60μm thick) [33]

Feed concentration / (wt%)	T / °C	Water flux / (kg m^{-2} h^{-1})	Selectivity
85	70	2.00	450
95	70	0.47	3953

Full-IPNs have higher compressional strength (2929 g load for 50 % compression) than the corresponding semi-IPNs (1883 g load for 50 % compression) [42].

These membranes can swell in water and different aqueous solutions evidencing the following aspects:

- the presence of PAAm affects in a positive way the swelling;
- the swelling in water increases with temperature (positive thermosensitivity);
- -swelling in the water/ethanol mixture increases linearly with the water content.

Because of membrane preferential swelling in different aqueous solutions, it may be recommended for use in separation processes by pervaporation.

The PVA/PAAm IPN membranes were found to have pervaporation separation factors ranging from 45 to 4100 and permeation rates of about 0.06-0.1 kg m^{-2} h^{-1}, for 95 % ethanol aqueous solution, at 75 °C [46]. For a concentration of 10 wt% ethanol, the permeation rates were as large as 9 kg m^{-2} h^{-1} and the separation factors were about 20 [46].

Recently, a new effective membrane for dehydration of different organic solvents by pervaporation has been reported. Novel hydrophylic polymer membranes based on crosslinked poly(allylamine hydrochloride) (PAA.HCl)-PVA have been developed [33]. The crosslinking agent was glutaraldehyde (GA). The role of PVA into the membrane is to increase its flexibility and the stability. But the increasing of PVA percentage, determines the decreasing of the water selective aminehydrochloride functional groups amount and as consequence, the rate of water intake by the membrane decreases. So, for different specific applications, the optimization of the PAA.HCl/PVA ratio in the formulation is essential. Also, the amount of GA and curing temperature has to be optimized to obtain the desired membrane properties. The characteristics of the ethanol dehydration process, by the pervaporation technique, are presented in table 4.

Polymers, such as polysaccharides (cellulose and chitosan (CS)) show a stronger affinity to water; hence their copolymers, blends or composites have been widely investigated for pervaporative (PV) separation of EtOH/H$_2$O mixtures [58-60]. Chitosan is generally preferred due to its high abundance, natural occurrence, hydrophilicity, chemical resistance, adequate mechanical strength, good membrane forming properties and ease of processing. PV performance of EtOH/H$_2$O mixtures through the surface crosslinked CS composite membranes exhibit a high selectivity value but a low permeation flux [61]. The PV membranes of derivatives of CS obtained by chemical modification have also been widely studied [62,63].

B.-B. Li et al. [64] have studied the separation of EtOH-H_2O solutions by pervaporation (PV) using chitosan (CS), poly (vinyl alcohol)-poly(acrylonitrile) (PVA–PAN) and chitosan-poly(vinyl alcohol)/poly(acrylonitrile) (CS–PVA/PAN) composite membranes. It was found that the separation factor of the CS–PVA/PAN composite membrane increased with an increase of PVA concentration in the CS–PVA polymer from 0 to 40 wt%. With an increase in the membrane thickness from 12 to 18 μm, the separation factor of the CS–PVA/PAN composite membrane increased and the permeation flux decreased. With an increase of ethanol–water solution temperature, the separation factor of the CS membrane decreased and the permeation flux of the CS membrane increased while the separation factor and the permeation flux of PVA/PAN and CS–PVA/PAN composite membranes increased.

Sodium alginate (SA), which is one of the polysaccharides extracted from seaweed, has shown excellent water solubility [65], but the mechanical weakness of SA membranes has been a drawback as a pervaporation membrane material. The use of SA–PVA blended membranes prepared by physical mixing of components, in different ratios, for pervaporation dehydration is reported elsewhere [66,67]. Taking on the basis of SA-PVA membranes, Dong et al. [68] studied the PVA–SA hollow-fiber composite membranes for organic dehydration by pervaporation. In particular, a polysulfone hollow-fiber membrane is coated by a PVA-SA blended solution. The founded optimal process of preparing membranes is as follows: 80 wt% PVA and 20 wt % SA are blended, and the casting solution of the PVA–SA blend with a concentration of 2 wt % is obtained by dissolving the blend in water; then the blend solution is cast onto the PS hollow-fiber membrane, and the composite membrane is crosslinked with 1.5 wt% maleic acid and 0.05 wt% H_2SO_4 in ethanol solvent for 8 h. For isoproanol, n-butanol, tert-butanol and ethanol aqueous solutions, as the alcohol concentration is 90 wt% at 45 °C, higher separation factors and permeation fluxes of crosslinked PVA–SA blended membranes are obtained: 1727, 414 g m^{-2} h^{-1}; 606, 585 g m^{-2} h^{-1}; 725, 370 g m^{-2} h^{-1} and 384, 384 g m^{-2} h^{-1}, respectively. This shows that these blended membranes have the potential to be used in industry.

2.1.7. Acetic Acid/Water Separation by Pervaporation

Poly(vinyl alcohol) and polyacrylamide (PAAM) blends, obtained by the different methods described above, can also be used for acetic acid dehydration, due to its capacity to swell in mixtures of acetic acid/water.

Swelling in water/acetic acid mixture shows a maximum of swelling shifting to higher temperatures when higher acetic acid concentrations increase (from 20 °C for 50 % acetic acid to 40 °C for 70 %).

Water is preferentially sorbed by membranes, but much less from water-acetic acid mixtures than from ethanol/water mixtures [46].

Table 5 presents the characteristics of the pervaporation process.

Table 5. Characteristics of the separation process by pervaporation according to membrane composition and structure, composition of feed mixture and temperature [18,52]

Compositi-on of the membrane PVA/PAcr.Ac	Composition of liquid mixture		Permeation rate / / (g m^{-2} h^{-1})	T / °C	Separation factor
75/25	Acetic acid-water	90/10	5.6	30	795

A recent paper [69] presents a new type of PVA hybrid membrane prepared by hydrolysis followed by condensation of a PVA and a tetraethylorthosilicate (TEOS) mixture, which shows a significant performace in water-acetic acid mixture separation. The highest separation selectivity (1116) with a flux of 3.33×10^{-2} kg m^{-2} h^{-1} at 30 °C for 10% mass of water in the feed has ben obtained by using the membrane containing 1:2 mass ratio of PVA and TEOS. The performance of these membranes was explained on the basis of a reduction of free volume and a decrease of the hydrophylic character owning to the formation of covalently bonded crosslinks. Significant lower apparent activation energy values have been obtained for water permeation comparatively to these of acetic acid permeation. The close values obtained for activation energy for total permeation and water permeation signify that the coupled transport is minimal due to the selective nature of membranes. The equal magnitude of activation energy for water permeation and activation energy for water diffusion indicates that both diffusion and permeation contribute almost equally to the PV process. The Langmuir mode of sorption dominates the process for all types of studied membranes.

Another recent work presents the possibility to use a membrane made by PVA-g- acrylonitrile (AN) to separate acetic acid/water mixtures by pervaporation [70]. The best separation factor (14.6) has been obtained by using PVA-g-AN (52 %) membrane, at 30°C, 90 % acetic acid in the feed. The permeation rate was 0.09 kg m^{-2} h^{-1}.

2.1.5. Separation Caprolactam (CPL)/Water Mixtures by Pervaporation

Caprolactam (CPL) is the monomer of *Nylon-6*, extensively used in high quality *Nylon-6* fibers and resin obtaining. Worldwild capacities reached above 4.5 million metric tones in 2005. A CPL dehydration study has been performed by pervaporation, using PVA crosslinked membranes (with GA as crosslinker agent and heat treatment of the membrane) [71]. In spite of the excellent dehydration performance for CPL/water mixtures exhibited by PVA crosslinked membranes (total permeation flux by 800 g m^{-2} h^{-1} and separation factor by 575, for PVA membrane crosslinked with 0.5 wt% GA, at 323 K and 50 wt% CPL in the feed), the authors recommended the use of a composite membrane with an active layer made by PVA, due to the poor durability and mechanical strength of the studied membrane.

2.1.6. Separation of Fluoroethanol/Water Mixtures by Pervaporation

2,2,2,-trifluoroalcohol (TFEA) is used for obtaining 2,2,2-trifluoroethyl methacrylate (TFEMA), necessary for preparation of functional water repellent paints and optical fiber coating agents. TFEMA can be manufactured by esterification of TFEA and methacrylic acid (MA) in the presence of an acid catalyst, at 70 °C. To obtain a higher conversion rate it is necessary to remove the water from the system, avoiding the formation of the thermodynamic equillibrium composition.

To attain this goal, a pervaporation technique has been proposed, using a PVA composite membrane, made by casting of a mixture of PVA aqueous solution and a GA one on a polyethersulfone (PES) porous support, solvent evaporation and thermic curing [72]. Excellent dehydration performance has been obtained (separation factor 320 and permeation flux 1.5 kg m^{-2} h^{-1}, for 90 wt% TFEA in the feed and 80 °C).

2.1.7. Separation of Methacrylic Acid/Water Mixtures by Pervaporation

A PVA composite membrane, made by casting of a mixture of PVA aqueous solution and a GA one on a polyethersulfone (PES) porous support, solvent evaporation and thermic curing, has been used to attain this aim [72]. Excellent dehydration performance has been obtained (separation factor 740 and permeation flux 2.3 kg m^{-2} h^{-1}, for 90 wt% TFEA in the feed, and 80 °C).

2.1.8. Water Desalination

PVA/Poly(ethylene glycol) (PEG) membranes crosslinked by aldehydes and sodium salts were used in water desalination by pervaporation. The desalination of 8 % NaCl solution by pervaporation at 55 °C and 5.00 kPa (downstream pressure) resulted in a single stage salt rejection of 99% and the water flux of 14 kg h^{-1} [73].

2.1.9. Dehydration of Methanol by Pervaporation

Another important application of membrane-based pervaporation is a well-established and commercially exploited method for the dehydration of organic solvents; in particular the dehydration of alcohols is done with the help of high permselective (hydrophilic) poly(vinyl alcohol)/polyacrylonitrile (PVA/PAN) thin film composite membranes, under the trade name of "GFT- Gesellshaft Fur Trenntechnik" membranes. One of the key successes of PV is that, if suitable membranes can be produced with a high permeability and a good selectivity to water, it is possible to achieve an excellent separation, particularly at the azeotropic composition. However, more number of novel polymeric membranes are needed for a successful operation of the process in view of the fact that PV is environmentally cleaner than the conventional distillation; moreover, this process is energy intensive. Consequently the success of any membrane depends on a high flux, a good separation factor (selectivity) and a long-term stability as well as a favourable mechanical strength to withstand the cyclic modes of PV operating conditions, as described before.

Also, membranes from blends of PVA/Poly(acrylic acid) [PAcr.Ac.] show a selective permeability against different components of a liquid mixture. This property of membranes makes them useful for the separation of components from liquid mixtures by the pervaporation method, i.e., for methanol dehydration.

Recently, a novel hydrophylic polymer membrane based on poly(allylamine hydrochloride) (PAA.HCl)/PVA, crosslinked with GA, has been also tested for methanol dehydration by pervaporation technique [33]. Even if the reported results show a small selectivity of the last type of membrane, the blend's composition, the curing degree and the process conditions (temperature, feed concentration, etc.) could be used to obtain a better separation of methanol.

Table 6 presents a summary of the published results.

Table 6. Characteristics of the separation process by pervaporation according to membrane composition and structure, composition of feed mixture and temperature [18]

Composition of the membrane	Composition of liquid mixture		Permeation rate / / (g m^{-2} h^{-1})	T / °C	Separation factor	Ref.
PVA/PAcr.Ac : 80/20	Methanol-water	70/30	70	50	55	42
			340	70	28	
		90/10	109	70	465	
		95/5	33	70	2650	
PAA.HCl/PVA/GA 60/35/5	Methanol-water (%$_{wt.}$)	86.5/13.5	1800	60	23	33

Table 7. Dehydration of acetone, using membranes PAA.HCl (60%$_{wt.}$)-PVA (35%$_{wt.}$-GA(5%$_{wt.}$) (aprox. 60μm thick) [33]

Feed concentration / (% wt.)	T / °C	Water flux / (kg m^{-2} h^{-1})	Selectivity
86	50	1.80	2270

2.1.10. Dehydration of Acetone by Pervaporation

Novel hydrophilic polymer membranes based on crosslinked poly(allylamine hydrochloride) (PAA.HCl)-PVA have been developed in order to dehydrate different organic compounds by pervaporation [33]. The characteristics of the acetone dehydration process, using a pervaporation technique, are presented in the table 7. The high selectivity of the membrane should be noted. The selectivity and flux characteristics of these membranes are excellent compared with most of the known membranes.

2.1.11. Pervaporation of Ethanol/Toluene

PVA-PAcr.Ac. membranes have been tested also for ethanol separation from ethanol/toluene mixture, by using pervaporation technique. The reported data concerning the separation process characteristics are presented in table 8.

Table 8. Characteristics of the separation process by pervaporation according to membrane composition and structure, composition of feed mixture and temperature [18, 40]

Composition of the membrane PVA/PAcr.Ac	Composition of liquid mixture		Permeation rate / (g m^{-2} h^{-1})	T / °C	Separation factor
10/90	Ethanol/toluene	10-90% ethanol	25-480	30	300-80

2.1.12. Pervaporation of Ethanol/Benzene

PVA-PAcr.Ac.blends membranes are suitable also for separation of components in ethanol/benzene mixtures. Reported data are presented in table 9.

2.1.13. Pervaporation of Methanol/Toluene

Methanol/toluene mixtures could be separated by pervaporation technique using PVA/PAcr.Ac. blend membranes. Reported data are presented in table 10.

2.1.14. Separation Methyltertbutyl Ether (MTBE)/Methanol Mixtures by Pervaporation

MTBE is a well known enhancer of the number of octanes in gasoline and as excellent oxygenated fuel additives that decrease carbon monoxide emissions.

Therefore, MTBE has been one of the fastest growing chemicals of the past decade. MTBE is produced by reacting methanol with isobutylene from mixed-C4 stream liquid phase over a strong acid ion-exchange resin as catalyst. An excess of methanol is used in order to improve the reaction conversion. This excess has to be separated from the final product. The pervaporation technique, more energy efficient and with lower cost process, has been proposed as alternative to distillation [74].

Table 9. Characteristics of the separation process by pervaporation according to membrane composition and structure, composition of feed mixture and temperature [18,46]

Composition of the membrane PVA/PAcr.Ac	Composition of liquid mixture ethanol/benzene	Permeation rate/ / (g m^{-2} h^{-1})	T / °C	Separation factor	Permeate activation energy E$_a$ / (kJ mol^{-1})
20/80	10/90	30	50	110	19.2
SIPN	90/10	560		3.5	
30/70	10/90	12	50	650	27.6
SIPN	90/10	460		1.9	
30/70	10/90	6	50	1100	31.4
SIPN	90/10	360		53	

Table 10. Characteristics of the separation process by pervaporation according to membrane composition and structure, composition of feed mixture and temperature [18,40]

Composition of the membrane PVA/PAcr.Ac	Composition of liquid mixture methanol/toluene	Permeation rate/ / (g m^{-2} h^{-1})	T / °C	Separation factor
10/90	10/90	120	30	460
	30/70	265	30	50

A membrane prepared by PVA blending with PAcr.Ac. in aqueous solution, casting, solvent evaporation and then crosslinking by heat treatment (at 150 °C), has been used.

The obtained results show that the prepared membranes are methanol selective, but the performance of these membranes (separation factor=30, for PVA/Pacr.Ac.=80/20, 5 wt% methanol in the feed, 25 °C) is lower than those reported by J.W. Rhim and Y.K. Kim [75] (separation factor 1250 for PVA/Pacr.Ac.=75/25, 20 wt% methanol in the feed, 30 °C).

The authors suggested that a combination of pervaporation with a conventional separation technique such as a hybrid distillation-pervaporation system could be useful economically to break the azeotropy.

2.1.15. Pervaporation of Benzene/Cyclohexane

The separation of benzene/cyclohexane mixtures is one of the most important and most difficult processes. Cyclohexane is produced by catalytic hydrogenation of benzene. The unreacted benzene in the effluent stream must be removed for pure cyclohexane recovery. Separation of benzene and cyclohexane is difficult because they have close boiling points (difference only 0.6 K) and close molecular sizes [76]. It is generally thought that separating benzene/cyclohexane mixtures is mainly governed by solubility selectivity due to the interaction between benzene molecule and membrane. Hence, increasing benzene solubility in the membrane is essential to obtain high permselectivity toward benzene. Poly(vinyl alcohol) (PVA) is polar and hydrophilic, and is an ideal membrane material to separate benzene/cyclohexane mixtures [77]. The selection of PVA is also due to its economical cost, commercial availability and good membrane-forming properties. F. Peng et al. report [78] the synthesis of poly(vinyl alcohol) membranes incorporating crystalline flake graphite (CG-PVA membranes). These blends take advantage of structure of graphite being similar to that of benzene favouring, in this way, the adsorption and packing of benzene on graphite surface and, consequently, increase the selectivity. A CG-PVA membrane exhibits a higher separation factor of 100.1 with a flux of 90.7 g m^{-2} h^{-1} at 323 K for benzene/cyclohexane (50/50, w/w) mixtures, showing that the incorporation of graphite into the PVA matrix interfered the polymer chain packing and enhanced effectively fractional free volume, and thus favourable for components diffusing through the membrane. Another interesting approach reported by the some authors [79] is to perform pervaporation of benzene/cyclohexane by using β-cyclodextrin (β-CD)-filled cross-linked poly(vinyl alcohol) (PVA) membranes (β-CD/PVA/GA). In the present case, the very important properties of the β-CD are used to increase the perselectivity toward benzene. The permeation flux of β-CD/PVA/GA membranes increased when the β-CD content was 0–8 wt%, but permeation flux decreased slightly when the β-CD content was 8–20 wt%. The separation factor towards benzene increased when β-CD content was in the range 0–10 wt% and decreased slightly when the β-CD content was 10–20 wt%.

Compared with the β-CDfree PVA/GA membrane, the separation factor of the β-CD/PVA/GA membrane for benzene to cyclohexane considerably increased from 16.7 to 27.0, and the permeation flux of benzene increased from 23.1 to 30.9 g m^{-2} h^{-1} for benzene/cyclohexane (50/50, wt) mixtures at 323 K.

To solve the tradeoff between permeability and selectivity of polymeric membranes, organic-inorganic hybrid membranes composed of poly(vinyl alcohol) (PVA) and ɣ-glycidyl oxypropyl trimethoxysilane (GPTMS) were prepared by an in situ sol-gel approach for pervaporative separation of benzene/cyclohexane mixtures [80]. The permeation flux of benzene increased from 20.3 g m^{-2} h^{-1} for pure PVA membrane to 137.1 g m^{-2} h^{-1} for PVA-GPTMS membrane with 28 wt % GPTMS content, while the separation factor increased from 9.6 to 46.9, simultaneously. The enhanced and unusual pervaporation properties were attributed to the increase in the size and number of both network pores and aggregate pores, and the elongation of the length of the diffusion path in PVA-GPTMS hybrid membranes.

Another hybrid membrane was prepared by filling carbon graphite (CG) into poly (vinyl alcohol) (PVA) and chitosan (CS) blending mixture [81]. This blend membrane shows homogenous distribution of graphite particles, considerable alteration of hydrogen bonding interaction, remarkable decrease of crystallinity degree, dramatic enhancement of mechanical properties and significant increase of free volume in CG-PVA/CS, which may contribute for improving the separation performance of the membranes by the synergistic effect of blending and filling. Comparing the performance of this blend with that used for PVA and PVA/chitosan membranes, for C_6H_6/C_6H_{12} separation, that new hybrid membrane exhibits a highest separation factor of 59.8 with a permeation flux of 124.2 g m^{-2} h^{-1} at 323 K, 1 kPa.

2.1.17. Separation Cyclohexene/Cyclohexan Mixtures by Pervaporation

Solid PVA-Co^{2+} composite asymetric membranes have been prepared starting from PVA and two different salts: $Co(NO_3)_2$ and $Co(CH_3COO)_2$, respectively, in order to separate cyclohexene/cyclohexan mixtures. A facititated transport mechanism has been evidenced, due to the capacity of Co^{2+} ions to coordinate the olefin molecules [82]. The authors reported stronger complexation of Co^{2+} ions with cyclohexene in the case of PVA/ $Co(CH_3COO)_2$ mixtures then in the case of PVA/ $Co(NO_3)_2$ mixtures. It was found that for a concentration ratio of ($[Co^{2+}]/[OH]$) by 0.75 mol/mol, the permeation flux of PVA membrane containing Co^{2+} increases 2-3 times and the separation factor increses 50 times compared with pure PVA membrane.

2.1.17. Fusel Oil Components Separation

One of the main products of sugar manufacturing is molasses, which contains approximately 50% sucrose and 50% other components (water, various other organic components and inorganic salts). Because of its high sucrose content, a substantial portion of the molasses is used for the production of ethyl alcohol through fermentation. The by-products of the fermentation broths, more volatile than the alcohol, are mainly aldehydes with acetaldehyde being the principal component. The aldehyde is removed, as a distillation head product. The other by-product of the distillation step, the bottom product, is fusel oil. It is composed of several alcohols, primarily C3, C4 and C5 aliphatic alcohols. The separation of its components, using pervaporation technique and PVA/PAcr.Ac. blend as membrane has been reported [55]. The characteristics of the pervaporation process are presented in table 11.

Table 11. Characteristics of the separation process by pervaporation according to membrane composition and structure, composition of feed mixture and temperature [18,55]

Composition of the membrane PVA/PAcr.Ac	Composition of liquid mixture: fusel oil	Permeation rate/ / (g $m^{-2} h^{-1}$)	T / °C	Separation factor	Permeate activation energy E_a / (kJ mol^{-1})
90/10	Alcohols mixture with 10-30 % of water	5000	60	10	49.4-41.7 (water) 60.8-55.7 (EtOH)

2.2. SEPARATION BY EVAPOMEATION

Evapomeation is a new membrane-separation technique for liquids mixtures, which eliminates some disadvantages of the pervaporation technique such as the decreasing of membrane permselectivity, due to its swelling by the direct contact with the feed solution. In evapomeation technique the membrane is not in direct contact with the feed solution, only with the solution's vapors. In this way the swelling of the membrane could be suppressed and consequently, the permeation rates in evapomeation are smaller than those in pervaporation, but the separation factor is greater [83].

The differences between the pervaporation and evapomeation processes may be seen in figure 5.

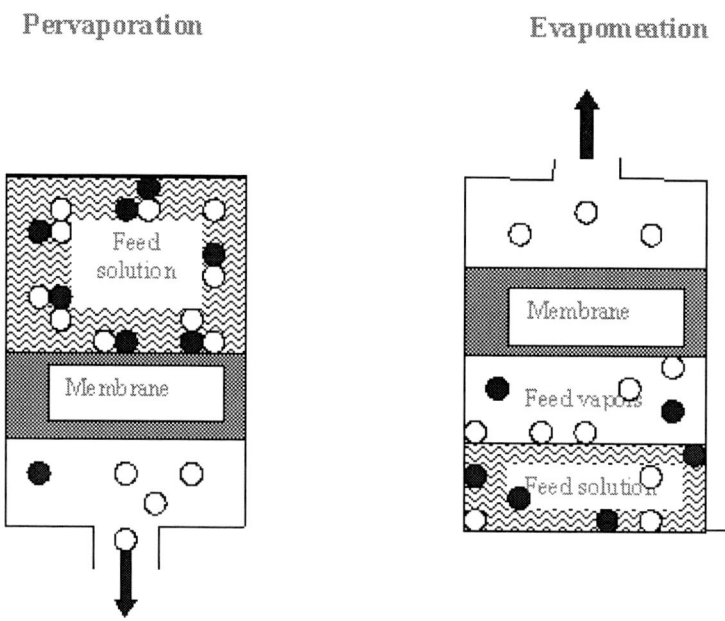

Figure 5. Schematic presentation of the pervaporation and evapomeation processes.

2.2.1. Isomers Separation by Evapomeation

Separation of n-propanol from a mixture of n-propanol (n-PrOH) and i-propanol (i-PrOH)

Taking into account the capacity of cyclodextrins (CD-s) to entrap a large number of organic and inorganic molecules, due to their hydrophobic cavity, β-CD has been introduced into PVA in order to obtain a good separation of isomer mixtures.

Two methods for obtaining PVA/(β-cyclodextrins) (β-CD) blends have been reported:

I. membranes were prepared by casting the solution (4%) of PVA (PD=1650; saponification degree = 99.7 %) and β-CD in DMSO at 25 °C and solvent evaporation at 80 °C [83,84];
II. by casting the aqueous solution of PVA (M_n=125,000), β-CD and 0.1% glutaraldehyde and water evaporation at room temperature in a vacuum oven for 24 h [85].

PVA and β-CD evidenced a good compatibility and produce transparent blend films [84].

The blend membranes are permselective for different organic isomers. So, these could be used for the separation of n-propanol from a mixture of n-propanol (n-PrOH) and i-propanol (i-PrOH) [84] and the separation of p-xylene from a p-xylene and o-xylene mixture [35]. It was evidenced that, in both cases, the separation was better by applying the evapomeation technique than that of the pervaporation.

It was observed that the n-PrOH concentration in the permeate through the CD/PVA membrane by pervaporation was approximately same as that in the feed solution, namely, the PrOH isomers could hardly be separated through these membranes by pervaporation. The n-PrOH concentration in the permeate obtained by evapomeation was higher than that in the feed solution, evidencing a higher permeation of the CD/PVA membrane for n-PrOH compared to i-PrOH.

The same situation was evidenced in the case of xylene isomers separation. The evapomeation is more efficient than that of pervaporation [83].

The n-PrOH concentration in the permeate and the normalized permeation rate increased with the increasing CD content in the CD/PVA membrane. The addition of CD in the PVA membrane determined the increasing of the swelling degree and preferential sorption of n-PrOH and p-xylene, due to the fact that the affinity of CD for these isomers was stronger than that for i-PrOH and o-xylene respectively [84].

The influence of the CD content in the membrane and the n-PrOH respectively p-xylene content in the feed mixture on the separation factors and sorption and diffusion selectivities of the CD/PVA membranes for the n-PrOH/I-PrOH and p-xylene and o-xylene mixtures by evapomeation are presented in tables 12 and 13.

It may be seen that a very high separation factor of organic liquid isomers through polymer membranes has been obtained for PrOH isomers [84].

A similar situation is reported for the separation of xylene isomers [83].

These results show the CD/PVA membranes are good candidates for isomers separation from organic liquid mixtures by evapomeation.

Table 12. Separation factors and sorption and diffusion selectivities of the CD/PVA and PVA membrane for the n-PrOH/i-PrOH (50/50 w/w) mixture and p-xylene and o-xylene (10/90 w/w) mixture by evapomeation versus the CD content [18, 83, 84]

CD content / / wt%	Separation of n-PrOH/i-PrOH (50/50 w/w) mixture			Separation of p-xylene and o-xylene (10/90 w/w) mixture		
	$\alpha_{sep.}$	$\alpha_{sorp.}$	$\alpha_{diff.}$	$\alpha_{sep.}$	$\alpha_{sorp.}$	$\alpha_{diff.}$
0	2.01	1.89	1.06	1.72	0.53	3.27
20	-	-	-	1.19	0.94	1.27
30	-	-	-	2.93	1.08	2.74
40	2.61	2.07	1.26	3.93	0.78	5.04

Table 13. Separation factors and sorption and diffusion selectivities of the CD/PVA (CD content: 40 wt %) membrane for the n-PrOH/I-PrOH mixture and p-xylene and o-xylene mixture by evapomeation versus the n-PrOH and respectively p-xylene, concentration in the feed [83,84]

Content of feed / / wt%		$\alpha_{sep.}$	$\alpha_{sorp.}$	$\alpha_{diff.}$
n-PrOH	p-xylene			
10	-	15.2	3.68	4.14
50	-	2.61	2.07	1.26
-	10	3.93	0.78	5.04
-	30	3.26	1.86	1.75
-	50	1.86	1.06	1.75
-	70	1.99	1.48	1.34
-	90	0.63	1.03	0.61

PVA/CD hydrogels swell in water. The swellability of PVA/CD hydrogels is marginally higher than that of the PVA gel, indicating that the crosslink density is higher in the PVA/CD system than in the PVA gel. The higher crosslink density may be an additional factor in retarding the migration of the drug in the presence of CD.

2.3. SEPARATION BY PERTRACTION

Pertraction is a continuous membrane-based extraction process, which has been proposed for, e.g., removing metal and organic pollutants from waste water treatment [86] and for concentrating valuable components from complex broths in bioproduction [87] as a consequence of solvent extraction. In this technology, the membrane contactor combines two functions, i.e., separation and extraction. It generally consists of a hydrophobic liquid phase so that the extraction and stripping of the solutes occurs in a three-phase system with two liquid/liquid interfaces. To this purpose, different techniques, such as impregnation of a microfiltration membrane by a circulating hydrophobic phase, supported liquid membrane and hydrophobic membrane, have been applied [88].

A very effective way to improve the pertraction performances in permeability and selectivity is to incorporate extractants into the hydrophobic phase, which react with a given solute reversibly and selectively.

S. Touil et al. [88] have reported the efficiency of membranes of cyclodextrin (CD)-containing PVA membranes (with CD covently grafted to the polymer chain) for the geometrical xylene isomer discrimination using the pertraction (combination of separation and extraction) technique. They found that in the presence of CD-containing membranes permeability coefficients of xylene isomers are higher when compared to control PVA membrane. It is also reported that α-CD is more effective to selectively extract the xylene isomers than β-CD. Flux observed for pertraction of single isomers and of the o-/p- binary mixtures was in the same order as the binding constants to α-CD i.e. p-xylene > m-xylene > o-xylene. The fabricated membranes exhibit a p-xylene selectivity for low p-xylene feed mole fraction (<70%) and a o-xylene selectivity for higher p-xylene feed mole fraction. The p-xylene enrichment factor observed at 10% p-xylene in the feed equal to 6, is the higher value ever reported for the separation of xylene isomers using CD containing membranes. The effect of way how CD, in particular α-CD, is introduced in PVA membranes is analysed in ref. 89. Physical trapping give rise to materials with an incorporation rate of 80 and 90% from the starting CD, whereas covalent attachment was quantitative. In both cases, however, it is found a high discrimination of p-and m-xylene over o-xylene.

2.4. SEPARATION BY MEMBRANE EXTRACTION

The development of membranes for, e.g., removing metal and organic pollutants from wastewater treatment [86, 90] and for concentrating valuable components from complex broths in bioproduction [87] is an attractive area.

Due to its structure and possibility to form complexes or inclusion compounds with different organic or inorganic substances, PVA hydrogels could be used to retain Cu(II) ions from waste waters (at pH values higher than 8). A green complex PVA-Cu (II) is obtained. The following equation of reaction could describe the PVA-Cu (II) complex formation:

$$\text{~CH}_2\text{-CH-CH}_2\text{-CH~} \atop \text{OH} \quad \text{OH} \;\; + \;\; Cu^{2+} \;\; \longrightarrow \;\; \text{PVA-Cu(II) complex}$$

This complex could be further used in S^{2-} ions retaining from wastewaters. Sulphide ions react with copper ions from the PVA matrix, leading to nanoparticles of CuS entrapped in the hydrogel. PVA-Cu (II) green hydrogel becomes black as it can be seen in figure 6.

The repartition constant of Cu (II) ions between the PVA cryogel and water has been determined as 13.45. The repartition constant of sulphide ions between the PVA-Cu(II) complex and water is very close of the first value, due to the high affinity S^{2-} to Cu^{2+} [92].

PVA hydrogel membranes obtained by freezing and thawing method could also retain iodine, in the presence of iodide ions. Red or blue complexes are formed in function of iodine concentration. A high repartition coefficient of iodine between cryogel and water has been obtained. They are dependent on the iodine concentration, evidencing a high level of interaction between PVA and iodine/iodide complex (K= 175 for 10^{-3} I_2/I^- aqueous solution and K= 455 for 3. 10^{-3} I_2/I^- aqueous solution) [93].

Taking into account that the iodine extraction from an aqueous solution is generally done by using very toxic and environmentaly dangerous organic solvents, such as CCl_4, $CHCl_3$, extraction of iodine with PVA hydrogel, non-toxic and biodegradable, could be a good candidate for "clean" technologies.

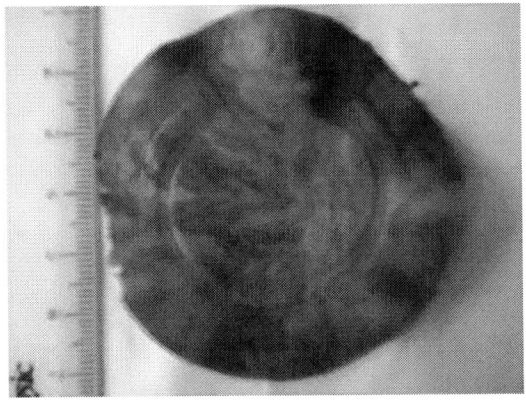

Figure 6. Photographic aspect of PVA-Cu (II) complex (a) and PVA-Cu(II)-CuS composite [91].

Another important task in environmental protection is nowadays the effective decontamination of medical wastewaters. The development of new medicines for the treatment of oncologic and chronic diseases brings with it the need to efficiently decontaminate media containing these medicines. The existing methods are intensely energy consuming or environmentally non-friendly (e.g., the use of ion exchangers with amine groups).

In the photodynamic therapy of cancer, for example, macrocyclic tetrapyrrholic compounds named porphyrins are used. Porphyrin-containing wastewaters can negatively affect the aquatic ecosystems (plants and fish population), even in very small concentrations. Recent studies present a method adequate for the advanced purification of medical wastewaters containing such porphyrins [94].

The method consists of the retention by sorption of the porphyrins on poly (vinyl alcohol) (PVA) hydrogels. Poly (vinyl alcohol) (PVA) is selected as the polymer of choice for the purification of industrial and medical wastewaters due to its capacity to form physically crosslinked hydrogels with the advantages of non-toxic, non-carcinogenic and biodegradable properties.

Some authors consider that poly(vinyl alcohol) hydrogels represent an efficient and environmentally viable alternative advanced purification method for porphyrin-containing medical wastewaters.

Many efforts to improve the efficiency and the selectivity of membrane processes are based on molecular recognition properties.

The incorporation of cyclodextrins in polymeric membranes will improve affinity and selectivity properties of those membranes due to the possible formation of host-guest complexes by supramolecular interactions. Cyclodextrins (CD) are cyclic oligosacharides having 6, 7, 8 or more glucose unities [95] called α-, β- and γ-CD, respectively. These compounds show a large number of applications once they exhibit complex formation with organic molecules, they are excellent models of enzymes which led to their use as catalysts, both in enzymatic and nonenzymatic reactions, and they are natural products and readily available. For these reasons, we can find applications of these compounds in analytical chemistry [96], drug carrier systems [97], etc. Cyclodextrins are water soluble macrocycles shaped like a rigid, truncated cone with a hydrophilic external surface and a relative non-polar cavity [98]. In fact, the cavity is lined by hydrogen atoms and glycosidic oxygen bridges. The non-bonding electron pairs of the glycosidic oxygen bridges are directed toward the inside of the cavity, producing a high electron density and lending it some Lewis base character. As a result of this special arrangement of the functional groups in the CD molecules, the cavity is relatively hydrophobic compared to water while the external faces are hydrophilic. These hydrophobic cavities provide an enormous host potential for molecular ability to form inclusion complexes with a large variety of organic and inorganic compounds in different solvents (including water) [99-101]. The selectivity originated from the different binding constants to CD depends on the size and shape of guest molecules. This fitting effect has been successfully exploited for separation of positional isomers and enantiomers in such techniques as HPLC and capillary electrophoresis [102,103]. It appears from literature that CD-containing membranes have been mainly based on the immobilization onto hydrophilic polymers acting as a barrier for hydrophobic compounds and thereby limiting their non selective diffusion [83,104,105].

Poly(vinyl alcohol) (PVA) seems to be one of the most efficient polymer matrix for CD-containing membranes owing to its ability to form free-standing films and its hydrophilic character due to the presence of hydroxyl groups. In these membrane materials CDs have been either trapped in PVA [83,84] or covalently linked to the chain [106].

Retaining of different ions from solutions is a very important target for wastewaters purification and for recovery of different ionic expensive species from solutions.

PVA is a non-ionic polymer, but it could be blended with ionic or ionizable polymers and it could be copolymerized or grafted, giving materials that exhibit ion-exchange capacity.

So, the PVA/poly(sodium styrene sulphonate) [PSSNa] blend was obtained by casting aqueous solution of polymers mixture (PVA with M_w= 124,000-186,000 and HD=99% and PSSNa with M_w= 70,000). The resulted films were crosslinked with 1,2-dibromethane in gaseous phase. A semi-interpenetrating network (SIPN) in which polyelectrolyte (PSSNa) chains are trapped inside a based PVA network was obtained [44]. A totally miscible blend with a very good film clarity and high mechanical resistance [44] resulted.

The membrane evidenced ion exchange capacity that depends on: crosslinking time (t_c) and the membrane composition. This capacity increases with the time of crosslinking from 0,8 to 2,0 meq/g after t_c= 2 respectively 12 h. The best result for ion exchange capacity was obtained for membranes with 45% PSSNa content. The membrane kept about 60% of the initial exchange capacity after more than 2 years [44].

The blends swelling ratio in pure water is shown as a decreasing function of crosslinking time [44].

The membrane, initial supple (t_c<2 h) becomes stiff and brittle after a longer crosslinking time (t_c =8 h).

The PVA/PSSNa membranes evidence a high permselectivity, comparable with the one of commercial ion exchange membrane as it can see in table 14, where were presented the permeability coefficient (P) and the ratio P to D (diffusion coefficient) that express the effect of porosity and of the electrolyte exclusion.

Table 14. Diffusion of sodium chloride at 25 °C through a PVA/PSSNa membrane[*] (Na^+ form) [18, 44]

C / (mol L^{-1})	0.01	0.1	1
P / ($cm^2 s^{-1}$)	1.91×10^{-7}	7.06×10^{-7}	2.50×10^{-6}
P/D	0.013	0.047	0.167

[*]Capacity: c_p=0.98 meq/g. Swelling ratio:τ_g=0.47. Thickness of dry and swollen sample: 140 and 200μm.

Also, the PVA/Poly(1,1 Dimethylenepiperidinium chloride) (PDMeDMPCl) blend membrane evidenced ion exchange capacity that increased with the time of crosslinking (t_c) from 0,92 to 1,2 meq/g after t_c= 15min respectively 120 min and showed a maximum capacity value, function of the weight fraction of PDMeDMPCl at 0.45 [44].

These membranes kept about 50% of the initial exchange capacity after more than 2 years.

The PVA/PDMeDMPCl blend membranes evidenced a lower permselectivity than PVA/PSSNa membranes, probably because of a possible phases' separation during the solvent evaporation, as it can see from table 15 [44].

Table 15. Diffusion of sodium chloride at 25 °C through a PVA/PDMeDMPCl membrane[*] (Cl^- form) [18,44]

C / (mol L^{-1})	0.01	0.1	1
P / ($cm^2 s^{-1}$)	2.38×10^{-6}	6.31×10^{-6}	7.17×10^{-6}
P/D	0.16	0.42	0.48

[*] Capacity: c_p=0.83 meq/g. Swelling ratio:τ_g=0.67. Thicknesses of dry and swollen sample: 80 and 150μm.

The PVA/PAcr. Ac. blends may also act like an ion exchange membranes if they are treated with 1,2-dibromoethane in gas phase. The average capacity of ion exchange is 6 mequivalent /g and depends on the weight fraction of the crosslinkable polymer [44].

Biosorption or bioaccumulation, the process of passive cations binding by dead or living biomass, represents a potentially consecutive way of removing toxic metals from industrial wastewaters. Biosorption could be employed most effectively in a concentration range below 100 mg L^{-1}, where other techniques are inactive or too expensive.

Metal ion binding during biosorption processes has been found to involve a complex mechanism, such as ion-exchange, complexation, electrostatic attraction or micro precipitation.

There have been some indications that ion-exchange plays an important role in metal sorption by algal biomass. Although numerous papers on the metal–microorganism interactions are available in the literature, still large uncertainties exist. Biosorbents are complex and variable materials. The composition of cell wall, to which metal ions are bound, depends not only on biosorbent species, but also on environmental conditions of its growth.

Recent studies confirmed that Azolla Caroliniana Wild fern, which is known as an effective bioacumulator in living state, is effective also in dry state (higher than 91% for Cr (III) ions retention) [107].

The dried fern particles have been also treated with nitric acid aiming to eliminate of cations initially present in the fern's body and to enhance its bioaccumulation capacity, but no important modification has been evidenced by this treatment [108].

The use of Azolla Caroliniana dry fern in water depollution avoids the problems of plant acclimatization in different climate conditions or polluted water characteristics and the water re-pollution by toxics delivery from died fern maintained in water.

The insertion of dried fern in a polymeric matrix avoids the fern particles mechanical degrading and permits the bioaccumulating material regeneration and it is reworking, determining the effectiveness of this advanced cleaning wastewater.

Taking into account the non-toxicity and biodegradability of PVA, this depollution method is an ecological one.

2.5. SEPARATION BY ULTRAFILTRATION

Ultrafiltration (UF) is a membrane separation technique used to separate extremely small particles and dissolved molecules in fluids, using suction or pressure. In membrane separation systems, liquid containing two or more components comes into contact with a membrane that permits some components (for example, water in the fluid) to pass through the membrane (the permeate), while other components cannot pass through it (the retentate). The physical and chemical nature of the membrane (e.g., pore size and pore distribution) affect the separation of the liquid and its components. The primary basis for separation is molecular size (normally higher than 15-200 Å), although other factors such as molecule shape and charge can also play a role. Molecules larger than the membrane pores (0.001 to 0.1 μm) will be retained at the surface of the membrane (not in the polymer matrix as they are retained in microporous membranes) and concentrated during the ultrafiltration process.

Compared to non-membrane processes (chromatography, dialysis, solvent extraction, or centrifugation), ultrafiltration is far gentler to the molecules being processed, does not require an organic extraction which may denature labile proteins, maintains the ionic and pH milieu, is fast and relatively inexpensive, can be performed at low temperatures (for example, in the cold room), and is very efficient and can simultaneously concentrate and purify molecules.

The retention properties of ultrafiltration membranes are expressed as Molecular Weight Cutoff (MWCO). This value refers to the approximate molecular weight (MW) of a dilute globular solute (i.e., a typical protein) which is 90% retained by the membrane. However, a molecule's shape can have a direct effect on its retention by a membrane. For example, linear molecules like DNA may find their way through pores that will retain a globular species of the same molecular weight.

There are three generic applications for ultrafiltration. a) Concentration: ultrafiltration is a very convenient method for the concentration of dilute protein or DNA/RNA samples. It is gentle (does not shear DNA as large as 100 Kb or cause loss of enzymatic activity in proteins) and is very efficient (usually over 90% recovery). b) Desalting and Buffer Exchange (Diafiltration): ultrafiltration provides a very convenient and efficient way to remove or exchange salts, remove detergents, separate free from bound molecules, remove low molecular weight materials, or rapidly change the ionic or pH environment. c) Fractionation: ultrafiltration will not accomplish a sharp separation of two molecules with similar molecular weights. The molecules to be separated should differ by at least one order of magnitude (10×) in size for effective separation. Fractionation using ultrafiltration is effective in applications such as the preparation of protein-free filtrates, separation of unbound or unincorporated label from DNA and protein samples, and the purification of PCR products from synthesis reactions.

Ultrafiltration (UF) is an important component in wastewater treatment and in food industry [109,110]. With increasing concerns and regulations in environment as well as in food safety, the process of ultrafiltration has become more critical, whereby new technology development to provide faster and more efficient water treatment is not only necessary but also urgent. Currently, conventional polymeric UF membranes are prepared mainly by the phase immersion process, typically generating an asymmetric porous structure with two major limitations: (1) relatively low porosity and (2) fairly broad pore-size distribution [111,112].

As a result, these membranes suffer two deficiencies: low flux rate due to the low porosity (i.e., limited permeability) and high fouling rate due to the asymmetric pore-size distribution having small pores on the surface [113].

In fact, the main problem with UF, however, is the flux decline caused by the irreversible adsorption of foulants onto the surface or even inside the pores of the membrane.

Solute adsorption often involves hydrophobic interactions—hydrophobic membranes have a high tendency to foul in water treatments. However, many hydrophobic membranes remain the most useful media for ultrafiltration due to their superior performance in terms of mechanical, chemical and thermal stability.

One approach to reduce fouling is using hydrophilic polymers, such as cellulose acetate (CA). Although CA membranes have outstanding properties in reducing membrane fouling, they lack long-term chemical, thermal and biological stability. Therefore, much research has focused on the development of good hydrophilic UF membranes using a high hydrophilic polymer.

To avoid, or diminish the fouling process, a blend of PVA with cellulose (CELL) has been obtained by spin coated of the PVA (M=50 000; 99% hydrolysis degree) / glutaraldehyde (GA) solution (corresponding to 0.005 or 0.01 moles of GA/ mole of PVA repeat unit) mixture onto the regenerated cellulose membranes (M=10 000) [114]. A composite structure has been obtained.

This hydrogel coating may penetrate the larger pores of the cellulose membrane and can exclude protein from entering them (100% protein retention). So, the hydrogel coating reduces the irreversible fouling of the cellulosic surface.

Relative water fluxes varied from 97.4 to 46.8 % as the thickness of the coating under hydrostatic pressure varied from 0.8 to 5.2 µm.

This blend is recommended as thin-gel composite membranes for bovine serum albumin ultra-filtration [114].

Polyvinyl alcohol (PVA) polymer is an attractive material to be developed as a new type of UF membrane with good anti-fouling characteristics. PVA membranes have a high level of mechanical strength, low fouling potential, longterm thermal resistance and pH stability. It has also been shown that PVA has good resistantce to most solvents besides strong polar solvents such as water, dimethyl acetamide, and N-methyl-2- pyrrolidone.

Many studies relative to PVA-based membrane materials focused on modifying commercial membranes to improve their anti-fouling performance [115,116]. For example, Na and Liu [117] reported that a PVA-based composite UF membrane could improve membrane hydrophilicity and its anti-fouling performance. The anti-fouling PVA composite membranes were dynamically prepared with an aqueous solution containing PVA, cross-linking agents and additives passed through porous substrate membranes such as polyacrylonitrile, polyvinylidene fluoride and Nylon.

X. Wang et al. [118] reported a new type of ultrafiltration membrane based on a different type of nanostructured porous support—electrospun nanofibrous scaffold in conjunction with a very thin layer of hydrogel coating to minimize fouling. Both the nanofibrous mid-layer support and the top coating layer were manufactured from crosslinked hydrophilic poly(vinyl alcohol) (PVA), where the degree of hydrolysis and the molecular weight of PVA were simultaneously adjusted to partially optimize the filtration performance and the mechanical durability. In that study PVA is the base material for fabrication of both the porous nanofibrous mid-layer support and the non-porous top coating layer. PVA is often used in ultrafiltration because of its superb hydrophilicity, biocompatibility, chemical and thermal stability. However, as PVA is water-soluble, it must be crosslinked to form water-resistant articles. PVA can be crosslinked through the reaction with hydroxyl groups using a wide range of chemicals [119,120]. X. Wang et al. synthesised a new high flux ultrafiltration nanofibrous composite membrane containing a crosslinked PVA electrospun scaffold and a PVA hydrogel coating.

The crosslinked electrospun scaffold using 96% hydrolyzed PVA with high molecular weight (85 000–124 000 g/mol) exhibits the best overall mechanical performance with high tensile, strength and elongation. The crosslinking reaction only resulted in a minor shrinkage in volume (<5%) in the electrospun scaffold, whereby the resulting porosity was relatively high (>80%). The PVA coating layer on the electrospun scaffold was crosslinked by using GA at varying concentrations.

Although the PVA hydrogel coating layer is macroscopically non-porous, it acts microscopically as a mesh of hydrophilic chains connected by crosslinking points. The mesh size can be controlled by the degree of crosslinking in the hydrogel and the best permeation rate and filtration efficiency is achieved by using the GA/PVA repeat unit ratio of 0.06 to crosslink the top PVA layer. The ultrafiltration test indicates that the flux rate of PVA nanofibrous composite membranes is at least several times better than those of existing thin film composite membranes [121-123], where its performance can be further optimized by reducing the top layer thickness or changing the layer composition.

Another way of using PVA for UF membranes is by modifying PVA by controlling hydroxyl groups. In this way the pore structure can be easily adjusted by the method phase inversion. Otherwise, once PVA is a water –soluble polymer it is difficult to form porous UF membranes with an ideal morphological structure by the method of wet phase inversion directly when water is used as a coagulation bath.

Acetalization of PVA is commercially used in the modification of PVA [124]. In general, formaldehyde, acetaldehyde and butyraldehyde have been used in acetalization of PVA. However, the hydrolyzing temperature of poly(vinyl formal) and poly(vinyl butyral) is higher than their deformation temperature. The hydrolysis and crosslinking for both poly(vinyl formal) and poly(vinyl butyral) membranes are difficult to perform directly below their deformation temperature. The crosslinking of the acetalized PVA with glutaraldehyde is easy to carry out without damaging the membrane shape and structure. In the Ref. 124 a number of hydrophilic UF membranes using the acetalized PVA is presented. The UF permeation tests were carried out using bovine serum albumin (BVA) solution as the feed instead of water. The modified PVA membrane exhibits a high level of water permeation along with good retention of BSA. It was found that the modified PVA UF membranes are hydrophilic and showed a good tendency dramatically to relieve protein fouling, thereby providing a better alternative to commercial UF membranes.

Chapter 3

OTHER DOMAINS OF MEMBRANES APPLICATION

3.1. FUEL CELLS

Ion conducting polymers containing strong acidic groups (e.g., sulfonic acid) are of interest for a broad range of applications, such as biosensors, chemical sensors, catalysts, actuators, ion-exchange membranes and polymer electrolyte membrane (PEM) fuel cells [125-127]. PEM fuel cells, in particular, are being investigated as replacements to current power sources used in transportation and portable electronics [128]. In this application, the ion conducting polymer or PEM serves as both a cell separator, separating the anode from the cathode, and an electrolyte, conducting protons from the anode to the cathode. Although there are a number of advantages to PEM fuel cells (e.g., renewable fuels, environmentally benign, high efficiencies), there are also a number of key shortcomings with current PEMs that hinder fuel cell efficiency. These shortcomings include low proton conductivity at higher temperatures, poor water management and high fuel crossover. Fuel crossover is a main concern as it applies to the methanol fuel-based PEM fuel cell (also known as the direct methanol fuel cell (DMFC)).

Direct-methanol fuel cells (DMFCs) have attracted considerable attention for certain mobile and portable applications, because of their high specific energy density, low poison emissions, easy fuel handling, and miniaturization [129,130]. However, the methanol permeation through electrolyte membranes (usually called methanol cross-over) in DMFCs still is one of the critical problems hindering the commercialization [131,132]. Nafion®, a poly-(perfluorosulfonic acid) membrane, is the major membrane used in polymer electrolyte membrane fuel cells (PEMFCs) presently. However, Nafion® membrane has a poor barrier property to methanol crossover (high methanol permeability). The methanol crossover to cathode not only reduces fuel efficiency, but also increases the overpotential of the cathode, thus resulting in lower cell performance [133]. The reason is known to be originated from protonic drag of methanol and diffusion through the hydrophilic channels within the membrane. Therefore, the effective methods for reducing methanol cross-over are to decrease the average diameter of ion-rich hydrophilic domains and increase the hydrophobicity of membrane surface.

To date, much effort has been undertaken to develop new alternatives. For example, sulfonated aromatic polymers, i.e., polymers with the sulfonic acid groups directly attached to the main chain or carrying short pendant side chains with terminal sulfonic acid units, attract increasing interest because of their chemical and thermal stability, and the ease of the sulfonation procedure. Some of the proposed polymers are sulfonated polysulfone (SPSU) [134] sulfonated poly(phenylene oxide) (SPPO) [135] sulfonated poly-(ether ether ketone) (SPEEK) [136] poly(phenylquinoxaline) (PPQ) [137] and poly(benzeneimidazole) (PBI) [138].

Poly(2-acrylamido-2-methyl-1-propanesulfonic acid)(PAMPS) was found to show higher proton conductivity than partially hydrated Nafion due to the sulfonic acid groups in its chemical structure [139]; consequently it can be chosen as a component for a new proton-conducting electrolyte membrane [140]. However, PAMPS, shows also some limitations as, for example, is highly water-soluble. Another key factor for the development of proton-conducting polymer electrolyte membranes is the water swelling. Extreme swelling causes a loss of the dimensional stability, while low swelling reduces proton conductivity because of low water absorption of the membranes. Cross-linking is an efficient means to limit the swelling, also yielding the dimensional and thermal stability of the membranes. [141-143]. Another consideration is alcohol cross-leaking, which is a key issue in the practical use of DMFC, but it can be controlled effectively by adjusting the cross-linking density of the prepared membranes. [144]. J. Qiao et al. [144] have synthesised a family of conducting polymer membranes of chemically modified poly(vinyl alcohol) - poly(2-acrylamido-2-methyl-1-propanesulfonic acid) (PVAPAMPS) prepared on the basis of a new concept of binary chemical cross-linking. It has been demonstrated that the excessive swelling of pristine PVA-PAMPS can be well controlled by chemical cross-linking using n-butylaldehyde/terephthalaldehyde, n-hexylaldehyde/terephthalaldehyde, and n-octylaldehyde/terephthalaldehyde as binary cross-linking agents. By changing the spacer length of the auxiliary crosslinkers, PVA-PAMPS membranes produce promising swelling characteristics and very good mechanical properties and flexibilities. The type and the amount of water absorbed by the chemically cross-linked PVA-PAMPS polymer blends are dependent not only on the sulfonic acid amount, but also on the spacer length of the CH_2 chain in the auxiliary crosslinkers and the cross-linker composition. The membranes show a larger sorption of nonfreezing water relative to freezing water. For a PVA-PAMPS of 1:1.5 in mass, with O5T5 as a binary cross-linking agent, a proton conductivity of 0.12 S cm^{-1} at 25 °C and of 0.098 S cm^{-1} at 5 °C is reported. The same authors [145] have mofidied PVA-PAMPS polymer blends by introducing a further polymer, the poly(vinylpyrrolidone) (PVP). The proton conductive polymer membranes PVA–PAMPS–PVP has the best proton conductivity of 0.088 S cm^{-1}, at 25 °C, for a polymer composition PVA:PAMPS:PVP of 1:1:0.5 in mass, which is comparable to commercially available Nafion117, and a methanol permeability of 6.1×10^{-7} cm^2 s^{-1}, one third of Nafion 117 methanol permeability (1.7×10^{-6} cm^2 s^{-1} [146]), at room temperature.

The use of a modified PVA membrane as a proton-exchange membrane is reported in the Ref. 147. The chemical structure of poly(vinyl alcohol) membranes is modified via sulfonation, using sulfophthalic acid (sPTA) as a sulfonating agent. The ion-exchange capacity (IEC), water uptake, methanol permeability and proton conductivity properties are evaluated for a set of sulfonated PVA membranes, with a variety of degrees of substitutions. The permeability of methanol and proton conductivity of these polymers are compared with those of Nafion115. The values of methanol permeability and proton conductivity obtained are 18.0×10^{-7} cm^2 s^{-1} and 0.112 S cm^{-1} respectively. Methanol permeability values of the membranes treated with 10% sPTA, at different cross-linking times, are around 5×10^{-7} cm^2 s^{-1}, and proton conductivity values of the sulfonated PVA membranes ranged between 0.024 and 0.035 S cm^{-1}.

The effect of annealing temperatures (65 – 250 °C) and blend composition of Nafion® 117, solution-cast Nafion®, poly(vinyl alcohol) (PVA) and Nafion®/PVAblend membranes for application to the direct methanol fuel cell is reported in [148]. These authors have found that a Nafion®/PVAblend membrane at 5 wt% PVA (annealed at 230 °C) show a similar proton conductivity of that found to Nafion® 117, but with a three times lower methanol permeability compared to Nafion® 117. They also found that for Nafion®/PVA (50 wt% PVA) blend membranes, the methanol permeability decreases by approximately one order of magnitude, whilst the proton conductivity remained relatively constant, with increasing annealing temperature. The Nafion®/PVA blend membrane at 5 wt% PVA and 230 °C annealing temperature had a similar proton conductivity, but three times lower methanol permeability compared to unannealed Nafion® 117 (benchmark in PEM fuel cells).

3.2. Sensors

The sensitivity of hydrogels to a large number of physical factors like temperature [149], electrical voltage [150], pH [151-153], concentration of organic compounds in water [154], and salt concentration [155] make them promising materials for a broad range of applications as microsensors [156] and microactuators [154] in MEMS devices.

The following principles for the pH value detection are used in sensors based on the swelling behavior of hydrogels: changes of the holographic diffraction wavelength in optical Bragg grating sensors [157], shifts of the resonance frequency of a quartz crystal microbalance in microgravimetric sensors [158], a bending of micromechanical bilayer cantilevers [153], as well as a deflection of silicon membranes in piezoresistive pressure sensors [159].

The swelling ability of pH-sensitive hydrogels depends on the functional acidic or basic groups at the polymer back- bone. Due to the dissociation of these groups and the influx of counterions, the concentration of ions in the hydrogel is higher than in the surrounding solution. This causes a difference in osmotic pressure and results in a solution flux into the hydrogel and, consequently, a swelling. The interaction and repulsion of charges along the polymer chain also lead to an increased swelling. Equilibrium of ionic gels occurs when the elastic restoring force of the polymer network balances the osmotic forces. During the swelling process hydroxide ions are transported into the neutral gel, while during the shrinkage protons diffuse into the gel and neutralize the negative charged acidic carboxylate groups. This ion diffusional flux induces an electrical potential difference that drives the electromigration of the ions in the direction opposite to that of the diffusion.

In so called "Donnan equilibrium" the diffusional flux of the ions in one direction is equal to the electromigrational flux in the opposite direction, resulting in a net zero mass transport and a net zero charge transport. The change of the electrical potential at the gel–solution interface is a function of the pH value of the surrounding solution. A Nernst–Planck equation coupled with the Poisson and the mechanical equilibrium equations can be used to describe the gel swelling/deswelling process [160]. Because the gel response is typically diffusion driven, the time response of the volume change approximately follows the square of the sample dimension. Scaling to micro-dimensions enhances the time response. Consequently, a reduction of the sample size improves the sensor performance. G. Gerlach et al. discuss the influence of the preparation conditions of hydrogel (poly(vinyl alcohol)/poly(acrylic acid) blend films on the sensitivity and response time of the chemical and pH sensors [161]. These authors have used swelling degree hydrogel properties as chemo-mechanical transducers for pH value variation. The hydrogel swelling leads to a bending of a thin silicon membrane and, by this, to an electrical output voltage of the sensor chip. The influence of the gel swelling/deswelling kinetics on the response time and long-term signal stability of proposed pH sensors leads to a signal drift. Such drift depends on the pH value of the ambient solution and is caused by the slow continuous change of the electrical potential at the gel–solution interface. The influence of previous gel swelling states can be minimized by a prolonged rinsing in de-ionized water after every measurement at high pH values. It is described that measurements in solutions with pH < 3 and large pH changes should be avoided in order to maintain sufficient sensor sensitivity for a long time. In order to achieve high signal reproducibility of pH sensors, a compensation of previous output signal values should be used.

Due to the chemical interactions between PVA and boric acid that lead to directly proportionally of the swollen hydrogel shrinking and the boric acid concentration, a sensor for this acid, difficult to determine by classical titration because of its weakness, has been proposed [162].

The development and applications of optical chemical sensors have grown rapidly. Among all sensors, optical pH sensors have received the most attention because of the importance of pH measurement in various scientific research and practical applications [163]. Optical pH sensors are based on pH-dependent changes of the absorbance or luminescence of certain indicator molecules immobilized on/in certain solid substrates. The immobilization of pH indicators to solid substrates is a key step in the development of optical pH sensors. Till now, there are three widely used methods for immobilization of pH indicators namely, adsorption or impregnation; covalent binding and entrapment. The adsorption and entrapment methods are relatively easy, but the leaking out of the indicators is a serious problem. The covalent binding method is relatively complicated and time-consuming, but very reliable since the indicators is not likely to leak out [164].

There have been many reports in which the immobilization method was covalent binding. In fact, many pH indicators used in above reports own at least one active amino or carboxyl group so that they can be covalently bound relatively easily to a solid substrate [165,166]. Kostov *et al.* had discussed the immobilizing process of Congo red, neutral red and phenol red to an activated diacetylcellulose membrane, and found that the indicator of phenol red was difficult to be immobilized via their method because of no active amino [167]. On the other hand, three factors that impact on the longterm stability should be considered, namely, the pH indicators themselves, the substrates and the linking bonds between the indicators and substrates. The commonly used ester linkage and acid-amide linkage are not very stable in acidic or alkaline aqueous conditions.

Polymeric pH indicators, phenolphthalein-formaldehyde (PPF) and *o*-cresolphthalein-formaldehyde (CPF) were synthesized with phenolphthalein and *o*-cresolphthalein reacted by formaldehyde under alkaline conditions, respectively. They can be immobilized in hydrolyzed cellulose diacetate membranes (HCDA) mainly due to macromolecular entrapment, and can be covalently bound to poly(vinyl alcohol) (PVA) via the considerable newly produced hydroxylmethyl groups [168,169]. Phenol red (phenolsulfonphthalein) and its derivatives are commonly used for pH determination.

Phenol red immobilized PVA membrane for an optical pH sensor is developed based on the same approach, since the molecular structure of phenol red is similar to that of phenolphthalein. Phenol red was first reacted with the formaldehyde to produce hydroxymethyl groups, and then it was attached to PVA membrane via the hydroxymethyl groups. The changes of spectra characteristics after immobilization, the ionic strength effects, response time, reproducibility and long-term stability of the sensor membrane are discussed by Z. Liu et al. [170].

Sol–gel-based biosensors have attracted an enormous scientific attention is the last decades [171-179]. Despite the volume of the published work, inherent drawbacks associated with the nature and the synthetic routes followed for the preparation of such gels still exist. These include cracking of the films, high concentration of methanol/ ethanol in the resulted sol, and the most important point regarding the development of amperometric-based biosensors, the lack of conductivity.

Constantinos G. Tsiafoulis et al. report the electrochemical behaviour of a composite film based on ferrocene intercalated $V_2O_5.nH_2O$ xerogel ($FeCp_2$–VXG) with photocrosslinkable polyvinyl alcohol with styrylpyridinium residues (PVA–SbQ), in order to be used as an electrocatalyst and host protein platform to develop an amperometric biosensor.

PVA–SbQ has been extensively used as a matrix for the immobilization of proteins [180-183]. The hydrophilicity of the polymer matrix, the mild conditions that are used during the immobilization and photopolymerization procedure make PVA–SbQ an effective support material for the immobilization of proteins.

Using glucose oxidase as a model enzyme, prospects of GOx–PVA–SbQ/FeCp2–VXG modified electrodes for further biosensor work in terms of working stability and storage stability, dynamic range, compatibility to proteins, applicability to near neutral pH, permeability and electrocatalytic activity are evaluated. Comparing with other xerogel based architectures, vanadium pentoxide xerogel shows to be superior in terms of conductivity and compatibility to enzymes. The proposed electrocatalyst provides about 20% increase of the sensitivity compared with the pure mediator, is compatible with biomolecules and its applicability over the useful pH range for most of the (bio)sensors applications indicates promise for further use.

Low-cost, disposable, SiO_2/Si_3N_4 chemical field effect transistor (ChemFET) microsensors have been fabricated for pH measurements and adapted to biochemical applications by using polyvinyl alcohol (PVA) enzymatic layers deposited and patterned either by dip-coating, or spin-coating and photolithographic techniques. Both processes have been compared for the development and optimization of a creatinine-sensitive enzymatic field effect transistor (Creatinine-EnFET). The Creatinine-EnFET has been characterized by linear detection properties (sensitivity around 30 mV/pCreatinine) on the [10–1 000 µmol L^{-1}] concentration range.(Creatinine-EnFET) [184]. Chronic end-stage kidney failure affects many patients in the world. Since its development in the 1960s, kidney dialysis has allowed a great number of patients to survive. So far, these techniques, and haemodialysis in particular, have been under constant development so that the quality of health care and the patients' life expectancy can be improved. To go further, dialysis efficiency must be known precisely, requiring a continuous monitoring of different chemical species concentrations into the blood: urea, creatinine as well as the H^+, K^+ and Na^+ ions. Creatinine is the end product of creatine metabolism in mammalian cells. Therefore, it is an important diagnostic substance in biological fluids. Creatinine can be used for the diagnosis of renal, thyroid and muscle function. It plays a major role in treatment with external dialysis. Normal range for plasma creatinine is 35–140 µmol L^{-1}. However it can reach concentrations higher than 1000 µmol L^{-1} in the case of kidney dysfunction. Thus, creatinine detection has to be developed in the [10 – 1000 µmol L^{-1}] concentration range for haemodialysis applications.

A highly sensitive amperometric biosensor for glutamate has been fabricated by immobilizing enzyme in a photo-crosslinkable polymer, polyvinyl alcohol bearing a styrylpyridinium (PVA-SbQ), membrane on a palladium deposited screen-printed carbon electrode is reported in Ref. 181. The polymer was previously reported to be suitable for fabrication of a thin enzyme membrane (about 1 mm thick) [185]. Enzyme can be immobilized in the PVA-SbQ matrix with high surface density and retain their functional characteristics to a large extent for several months upon repetition of wetting and drying [186]. Moreover, enzymes can be immobilized in this polymer using photolithography techniques [187], which can be adapted to mass production using ordinary screen printing or semiconductor-fabrication processes on a planar electrode [188].

Strong electrochemical interference from oxidizable species, such as ascorbic acid and uric acid, in the biological samples exposes a serious problem for the practical operation of amperometric biosensors with a working potential of 0.4 V or higher [189]. For example, electrochemical oxidation of ascorbic acid (AA) generates the dehydroascorbic acid (DAA), with the loss of two electrons and the consequent loss of hydrogen ions. One way to solve this problem is to modify the electrode surface with a permselective membrane. A variety of polymer membranes have been reported to be useful for eliminating interferents [190-192]. These polymers show permselective properties based on size exclusion (e.g. poly-l-lysine and poly (4-stryenesulfonate) membrane) and/or charge interaction between solutes and the membrane (e.g., Nafion).

Biosensors fabricated on the Nafion and polyion-modified palladium strips are reported by C.-J. Yuan [193]. They found that Nafion membrane is capable of eliminating the electrochemical interferences of oxidative species (ascorbic acid and uric acid) on the enzyme electrode. Furthermore, it can restricting the oxidized anionic interferent to adhere on its surface, thereby the fouling of the electrode was avoided. Notably, the stability of the proposed PVA-SbQ/GOD planar electrode is superior to the most commercially available membrane-covered electrodes which have a use life of about ten days only. Compared to the conventional three-dimensional electrodes the proposed planar electrode exhibits a similar long-term stability, but is smaller, more responsive and more versatile. The manufacturing processes used in the semiconductor industry can be adapted to produce these electrodes at a unit cost that is low enough to ensure cost-effectiveness.

The blend of PVA with PEG- modified glucose oxidase could be used as glucose sensor characterized by the linearity of calibration curve in the range of concentration by $5 \times 10^{-5} - 5 \times 10^{-3}$ mol glucose L^{-1} [194].

3.3. BIOCHEMICAL/MEDICAL APPLICATIONS

In the recent years, the scientists' position concerning the diseases treatments was completely changed. No longer is the treatment of specific diseases, such as diabetes, asthma, cardiac problems, osteoporosis, cancer etc. based only on conventional pharmaceutical formulation. Biology and medicine are being to reduce the problems of disease to problems of molecular science, and are creating new opportunities for treating and curing disease. Such advances are closely related with advances in biomaterials and are leading to a variety of approaches for relieving suffering and prolonging life [195].

An exponential increase of the biomaterials application could be noted in the last years.

R. Langer and N.A. Peppas reported the main domains of biomaterials application that could be schematically represented in figure 7.

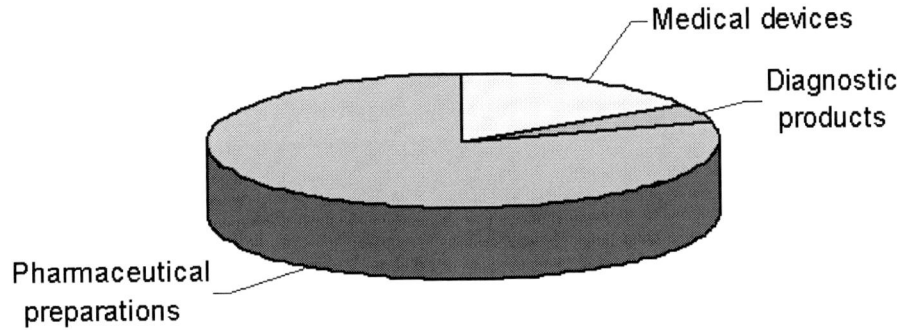

Figure 7. Repartition of the main domains of biomaterials applications.

The suitable materials for the above mentioned domains are polymers, metals and ceramics. Among these, polymers play an important role. Even the polymers have a lot of remarkable properties that could be used in biomaterials design, the interaction between these artificial materials and tissues and blood could create serious medical problems such as clot formation, activating of platelets, and occlusion of tubes for dialysis or vascular grafts. In the last few years, novel techniques of synthesis have been used to correlate desirable chemical, physical and biological properties of biomaterials.

One of the widely used categories of polymers for biomaterials design is that of homo-or copolymers, which could generate hydrogels. Hydrogels are three-dimensional polymer networks that could swell in water without dissolution and that, due to their high water content and rubbery nature, are very similar to natural tissues and could be considered biocompatible. Also, hydrogels may provide desirable protection for drugs, peptides and proteins from the potentially harsh environment in vicinity of the release site. Hydrogels could be also excellent candidates as biorecognizable biomaterials that could be used as bioadhesive systems, as targetable carriers of bioactive agents or as conjugates with desirable biological properties.

PVA is a well known polymer with large possibilities to be used as biomaterial, due to its non-toxicity, biocompatibility, non-carcinogenity, capability to react with other compounds and to be blend with a lot of polymers, changing its initial neutral network with positively or negatively charged one. Also, by modifying their initial structural characteristics such as molecular weight, hydrolysis degree, OH groups tacticity, some of PVA properties could be modified, such as mechanical and thermal resistance, water solubility, and chemical stability. Also their capability to be crosslinked by chemical or physical routs, to be blended with different polymers or copolymers, to be graft or copolymerize with different chemical partners lead to obtaining of intelligent hydrogels those stimuli-responsive properties could be relatively easy to tailor.

Also, PVA hydrogels evidenced a very good behaviour in contact with skin and other tissues, mucosa, or blood. PVA exhibits a bioadhesive nature, shape-memory properties, avoid the protein adsorption onto the gel surface and is biocompatible.

Recent reports [1,196] showed that PVA hydrogels are used as blood-compatible material, as contact lenses, as membranes for plasmapheresis, as artificial skin, as vocal cord reconstruction, as articular cartilage, in controlled drug delivery as neutral non-biodegradable matrix (in human body conditions), but more recent studies evidenced that by blending, by grafting or by copolymerization, by crosslinking by different methods, PVA-based hydrogels could be use also as temperature, pH, electrolyte-sensitive biomaterials.

Polyvinyl alcohol (PVA), which is a water soluble polyhidroxy polymer, is one of the widely used synthetic polymers for a variety of medical applications [197] because of easy preparation, excellent chemical resistance, and physical properties. [198] But it has poor stability in water because of its highly hydrophilic character. Therefore, to overcome this problem PVA should be insolubilized by copolymerization [43], grafting [199], crosslinking [200], and blending [201]. These processes may lead a decrease in the hydrophilic character of PVA. Because of this reason these processes should be carried out in the presence of hydrophilic polymers. Poly(vinyl pyrrolidone), PVP, is one of the hydrophilic, biocompatible polymer and it is used in many biomedical applications [202] and separation processes to increase the hydrophilic character of the blended polymeric materials [203,204]. An important factor in the development of new materials based on polymeric blends is the miscibility between the polymers in the mixture, because the degree of miscibility is directly related to the final properties of polymeric blends [205].

A very complete study, effect of pH, concentration of SA, PVA/PVP ratio and the temperature on the SA release, concerning the controlled delivery of SA from PVA/PVP membranes is given in the Ref. 206. Four main conclusion arise from this study: a) the presence of PVP increased the released amount of SA; suitable PVA/PVP ratio was found to be as 60/40 (v/v) for PVA/PVP membranes; b) the release percentage of SA through PVA/PVP membranes and swelling degrees of the PVP-40 membranes increase with an increase in the pH of donor solution; the pH of the acceptor solution do not affect much the transfer of SA through PVP-40 membranes; c) grafting of PVA with VP is more effective than blending with PVP for the release of SA; and d) the increase in the temperature increase the transfer of SA; the release percentage for SA is found being 57.5 % and 66.6 % at 32 °C and 37 °C, respectively.

Delivery of hydrophilic molecules such as proteins and DNA for therapeutic application is generally considered a great challenge [207,208], because these molecules are rapidly degraded by enzymes found under in vivo conditions both intracellularly at the site of application as well as in the general circulation, causing low bioavailabilies and requiring frequent injections [209]. Nanoscale carriers such as nanoparticles and nanocomplexes have reached increasing attention, since they can be administered by various routes, including the intravenous and intranasal routes [210,211]. Controlled and sustained release of these drug candidates can be accomplished using microspheres and implants from biodegradable polymers [207,212]. The classic copolyesters of lactic and glycolic acid (PLGA) are not ideal for protein and DNA delivery since inactivation and uncontrolled release is a consequence of poor compatibility between lipophilic polymers and hydrophilic drug candidates [213,214]. This is especially the case for DNA, where the complexation capabilities and protecting abilities of the carrier substance are very important. M. Wittmar et al. [215] selected PVA, once it is biocompatible and can be eliminated from the body by renal excretion [216,217]. To this polymer backbone, amine groups were covalently coupled in a polymer-analogous reaction using carbonyl diimidazole (CDI) to introduce cationic charges under physiological conditions [218,219]. This modification affects the colloidal stability of carrier systems by imparting positive surface charges on one hand [220] and increasing the protein or DNA loading by electrostatic interactions on the other hand [221,222]. As the secondary and tertiary amino-groups functions possess lower cytotoxicity, diamines, and PVA were coupled via the hydrolytically stable urethane bond [223]. The resulting PVA can be used in different ratios to complex DNA.

K.S. Oh et al. [224] have used PVA-containing matrices as temperature sensitive drug delivery systems. Their approach is based on the fact that the constant release is not the only way to accomplish the maximum drug effect and the minimum side effects and the assumption used for constant release rate sometimes fails its validity for physiological conditions. Such difficulty can be overcome by technology that senses environmental stimuli and appropriately controls the drug-release rate. Stimuli-sensitive polymers undergo phase transition in response to changes in, for example, pH, temperature, or the metabolites [225-227]. Especially, polymer materials with temperature-induced swelling transitions resulting from both polymer–water and polymer–polymer interactions have been reported [228]. K.S. Oh et al. have prepared a novel polymer complex gel composed of F-68 (Pluronic, poly(ethylene oxide)-poly(propylene oxide)-poly(ethylene oxide) triblock copolymer) and poly vinyl alcohol (PVA). The polymer complex gel if formed by intra/intermolecular interactions via hydrogen bonding in water. For the application as a temperature-sensitive delivery system of acetoaminophen, F-68/PVA complex gel is prepared with a form of polymeric bead encapsulated by poly(lactide-co-glycolide)(PLGA) membrane and pulsatile release of acetoaminophne, used as model drug, pattern is observed in response to pulsatile change of temperature between 35 ºC and 40 ºC.

A new material with good antithrombogenic properties, suitable as biomedical material which assures the endothelialization of the inner surface of a polyurethane tube to imitate the inner wall of a natural blood vessel has been synthesized by blending PVA with poly(carbonate urethane)(PCU) [229].

This blend was obtained by polymers mixture extrusion and extraction with the azeotropic mixture of hexane/ethanol, and modifying the obtained polymer surface by coupling of 4-isocyanato butanoic acid methyl ester (as a spacer molecule) to PVA blend, saponification of methyl ester groups and coupling of 4-amino-TEMPO (2,2,6,6-tetramethylpiperidine-1-oxyl) [229].

Generally is difficult to delimitate the medical or pharmaceutical application of PVA hydrogels as gel matrix, micro spheres, aerosols or membranes.

Taking into account the consensual accept of the membrane concept, we could consider as application in membrane form transerdmal patches, wound dressing, materials for tissue engineering, thin coatings with imprintig gels for molecular recognition, biomembranes in artificial organs, haemodialysis.

3.3.1. Transdermal Patches

The most common form of drug delivery is via the oral route. Although this has the notable advantage of easy administration, it also has significant drawbacks namely poor bioavailability due to hepatic metabolism and the tendency to produce rapid blood level spikes (both high and low), leading to a need for high or frequent dosing, which can be both cost prohibitive and inconvenient. Another method utilized in drug delivery is the systems that deliver the drugs through the skin into the bloodstream, making them easy to administer. In transdermal drug delivery, improved bioavailability, more uniform plasma levels, longer duration of action resulting in a reduction in dosing frequency, reduced side effects and improved therapy due to maintenance of plasma levels up to the end of the dosing interval, and patient compliance could be possible.

Transdermal patch technology represents an important area of biomaterials, due to its non-invasive character, ease to use, and a relatively high bioavailability. Generally, these patches could deliver drugs from one to seven days. Currently, 11 drugs, or drugs combinations are delivery through body via this method [195].

Nowadays, scientists are exploring various physical forces to enhance the transport through the skin to expend the number of drugs being delivery such as electricity, (iontophoresis, electroporation) or ultrasounds.

Drug delivery to the skin has been traditionally designed for dermatological drugs to treat skin diseases or for disinfection of the skin itself. In recent years, a transdermal route has been considered as a possible site for the systemic delivery of drugs. The possible benefits of transdermal drug delivery include that drugs can be delivered for a long duration at a constant rate, that drug delivery can be easily interrupted on demand by simply removing the devices, and that drugs can bypass hepatic first-pass metabolism. Furthermore, because of their high water content, swollen hydrogels can provide a better filling for the skin in comparison to conventional ointments and patches. Versatile hydrogels-based devices for transdermal delivery have been proposed [230].

Recently, porphyrins have been applied to cancer photodynamic therapy (also known as photochemotherapy), a method based on applying a porphyrinic compound onto the tumour and then irradiating with a light source. The porphyrin acts as a photosensitiser, transferring its energy to the oxygen found in tumoral tissue, generating singlet (radicalic) oxygen, which has the ability to oxidize tumour cells and also induce cell death (apoptosis). To stabilize and to assure a convenient delivery of porphyrins their entrapment in a hydrogel matrix has been proposed. Different porphyrins (figure 8), water soluble (5,10,15,20-tetra-sulphonato-phenyl porphyrin, TSPP) and water insoluble (5,10,15,20-tetra-pyridil porphyrin, TPyP, and 5,10,15,20-tetra-phenyl porphyrin, TPP)) have been immobilized in PVA cryogel matrix.

One can conclude that PVA hydrogels represent an efficient encapsulation vehicle for the studied porphyrins, both water soluble and non-water soluble. Their biocompatible, biodegradable, non-toxic, and non-carcinogenic nature makes them especially effective for pharmaceutical applications, but also for environmental uses, such as advanced wastewater decontamination. Hydrogels prepared from high molecular mass PVA have a better sorption profile for porphyrins, and are better suited for the preparation of controlled-release vehicles.

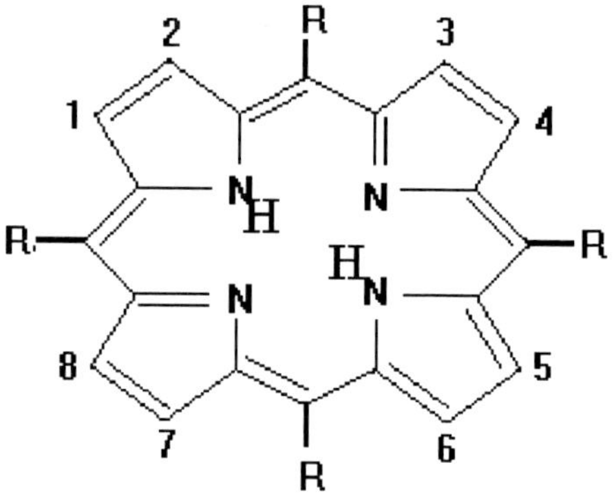

TPyP: R=-C$_5$NH$_4$; TPP: R=-C$_6$H$_5$;TSPP: R= -C$_6$H$_4$-SO$_3^-$Na$^+$

Figure 8. Structure of the porphyrins TPyP, TPP and TSPP.

The porphyrins structure and the PVA molecular weight determine differences in the porphyrins sorption onto the PVA hydrogels, as it can be seen in figure 9 [231].

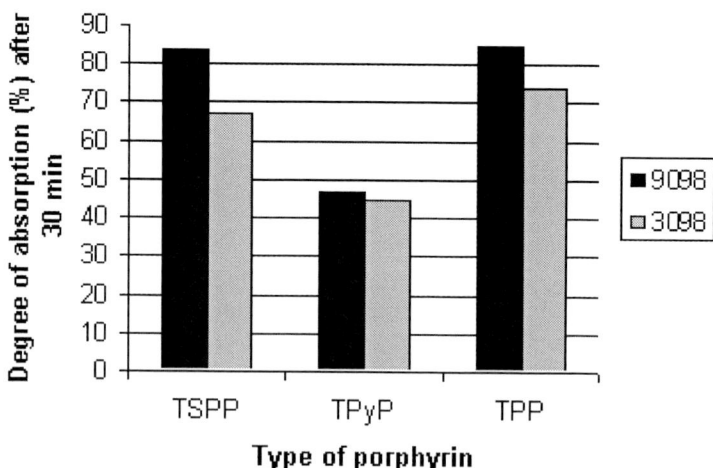

Figure 9. Comparison between the sorption degree of porphyrins on the PVA hydrogel, as function of PVA molecular weight and porphyrin type.

Also, PVA has a number of desirable characteristics that make it a good bioadhesive polymer. It has mechanical strenght, high elasticity, and swells upon immersion in water. Crosslinked PVA has been proposed as drug delivery carriers. In these gels the drug is able to be released fast or slowly due to the gel's high or low swelling ratio upon immersion in water. Previous studies by Morimoto et al. [232] have shown that several drugs such as indomethacin, glucose, insulin, heparin, and albumin can be released from crosslinked PVA gels. PVA properties depend upon the degrees of polymerization and hydrolysis. The solubility of PVA in water increases greatly as its degree of hydrolysis increases. Properties such as water solubility, high tensile strength and tack make PVA useful as an adhesive with fully hydrolyzed grades of PVA being water-resistant adhesives. PVA cryogel maximum adhesion has been achieved after two freezing/thawing cycles, saples prepared by three and four cycles still exhibit adhesive characteristics. The mucoadhesive and drugs release (i.e., theophyline and oxprenolol hydrochloride) behavior could be adjusted by degree of crystalinity which depend on the number of freezing/thawing cycles.

3.3.2. PVA Based Materials as Wound Dressing

The current generation of medical dressing differs from their predecessors primarily because they are based on non traditional polymers in this area. In creating new wound coverings, instead of cellulose stock, other natural and synthetic polymers are being increasingly used. One of them is PVA, because of its exceptional properties. Also, an important change in dressing form should be noted: in many cases, preference is given to granulated sorbents, hydrogels, films and sponges [233].

Applied to wounds, burns or surgical incisions, hydrogel materials cover the injured parts of skin (wounds) and promote healing and skin growth.

There are two ways of wound treatment: under dry or wet conditions. In the former gauze is applied to the skin, the latter calls for the use of hydrogels.

Since many agree that wounds heal faster in a wet environment, hydrocolloid type materials mixed with gelatin, hydrophobic polymers and water have been developed and already see practical use. However, these materials are mechanically weak and require periodic change, and the residue must be removed by washing with physiological salt solution. This process determines exfoliation of the new skin and delays healing. Also the removing of the dressing induces pain.

To avoid all these drawbacks new wound dressing materials have been tested. So, PEO/PVA hydrogel, crosslinked by electron beams, has been synthesized and tested as dressing on animals wounds. Wounds dressed with a hydrogel healed almost completely within 14 days, while those using gauze were only half-healed within that time. Clinical tests confirmed the safety and effects of this product [234].

The main features for a hydrogel dressing are the following: to protect injured skin and keep it appropriately moist to accelerate the healing process; to absorb liquids exuded by the body; to prevent infection from external bacilli; to be non-toxic, non-irritant and non-carcinogenic; to be soft and high adhering to the skin; to have mechanical resitance; to have high permeability; to resist to sterilizing process; to remove without pain; to be transparent to allow observation of the healing process.

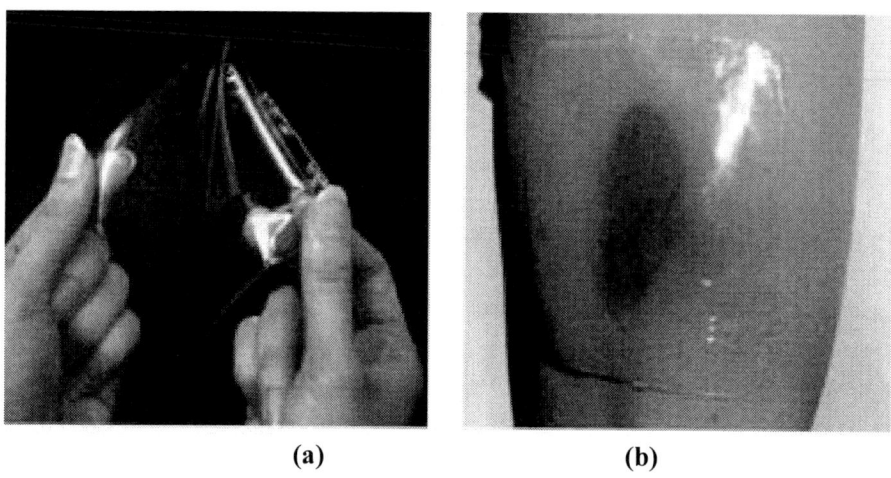

(a) (b)

Figure 10. Hydrogel membrane (a) and their application as wound dressing (b).

New generation of therapeutic coverings have to be characterized also by the ability to exercise a biologically active effect on wound. Incorporation of one or more drugs in the polymeric matrix that acts as a vehicle with controlled delivery capacity makes the new dressings able to exert anesthetizing, antimicrobial or combined effect. Also, the immobilization of proteolytic enzymes and an antimicrobial on a polymer substrate helps by decreasing the cleansing and healing time, especially in the purulent wounds treatment [235].

Table 16. Some examples of systems used as new wound dressings

Polymer system/ drug	Properties	Ref.
PVA/ proteolytic enzyme protease C/ polyhexamethyleneguanidine salt (PHMG) (antimicrobial factor (AM)) a/-PHMG (C)= PHMG hydrochloride; b/-PHMG (P)=PHMG phosphate	-PHMG influences the viscosity of the initial PVA solution (a higher decrease has been observed in case (a) then in case (b)) -(a) and (b) determines the decreasing of the activation energy of viscous flow -additive adding (sodium tetraborat as crosslinking agent) determines the increasing of the spinning PVA solution	233
PVA film/ Sodium tetraborate/ Proteolytic enzyme Protease C (Pr)+ polyhexamethyleneguanidine hydrochloride [PHMG], as antimicrobial (AM)	-the porous structure of the film and the repartition of the complex is influenced by the type of AM -biological active material -increases the AM desorption rateby 1.5-4.5 times For PHMG with M, 10000, total desorption of AM from the film could be obtained	236
PVA film/ Sodium alginate/ Proteolytic enzyme Protease C (Pr)+ polyhexamethyleneguanidine hydrochloride [PHMG], as antimicrobial (AM)	-biological active material -decreases the rate of inactivation of Pr by 2 times -decrease the amount of desorbed AM by 10 times, giving the film self-disinfecting properties.	236
PVA film/ Sodium alginate (Alg)/ Proteolytic enzyme Protease C (Pr)+, a cationic polymeric antimicrobial (AM=Metacid)	-AM-s are derivatives of PMGH obtained by neutralization with HCl (Metacid) and H_3PO_4 (Fogucid)	235
PVA film/ Tetraborate (TB)/ Proteolytic enzyme Protease C (Pr)+, a cationic polymeric antimicrobial (AM=Metacid)	-The interactions between AM-s and PVA determine the modifying of the composite film morphology and as consequence the modifying of the water and water vapors sorption (Metacid increases the films water sorption and Fogucid decreases it).	235
PVA film/ Sodium alginate/ Proteolytic enzyme Protease C (Pr)+, a cationic polymeric antimicrobial (AM=Fogucid)	-The films swelling capacity influence the AM-s delivery: Fogucid is desorbed more rapidly than Metacid.	235

Polymer system/ drug	Properties	Ref.
PVA film/ Tetraborate (TB)/ Proteolytic enzyme Protease C (Pr)+, a cationic polymeric antimicrobial (AM=Fogucid)	-Incorporation of additives also influence the AM-s desorption from the composite films: In both cases, the AM-s desorption is more difficult, due to the diffusion hindrances caused by the increasing of the diffusing particles as a result of formation of a complex, in the case of Alg., and caused by the intermolecular crosslinks, in case of TB. -The vapors water sorption rate gradually increases in time for pure PVA films and decreases as absolute values by comparing to PVA pure films when Alg or TB are added; a more important decrease could be obtained when AM with higher molecular weight is added.	235
PVA/β-CD/salicylic acid	-β-CD forms inclusion complexes with different water soluble substances i.e. salicylic acid - The drug release from the PVA/β-CD gel is nearly proportional to time	235
PEO/PVA	-wound dressing	234
PVA hydrogel UV crosslinked/ nitric oxide	NO release from NO-modified hydrogel occur over a time period up to 48 h., and there were no associated decrease in fibroblasts growth or viability in vitro associated with NO hydrogels. Exogeneous NO released from hydrogels wound dressing has potential to modulate healing.	237
acrylamide-functionalized nondegradable poly(vinyl alcohol) (PVA), UV- photo-crosslinked	-As the PVA content increased from 10% to 20%, protein flux decreased, with no trypsin inhibitor (TI) permeating through 20% PVA hydrogels; -Further increase in model drug release was achieved by incorporating hydrophilic PVA fillers into the hydrogel. As filler molecular weight increased, TI flux increased. - Release studies conducted using growth factor in vehicles with hydrophilic filler showed sustained release of platelet-derived growth factor (PDGF-β,β) for up to 3 days compared with less than 24 hours in the controls. In vitro bioactivity was demonstrated by doubling of normal human dermal fibroblast numbers when exposed to growth factor–loaded vehicle compared to control.	238

3.3. MATERIALS FOR TISSUE ENGINEERING

There are three ways in which materials have been shown to be useful in tissue engineering:

- the materials able to induce cellular migration or tissue recognition;
- the materials are used to encapsulate cells and act as an immunoisolation barrier;
- the materials are used as matrix to support cell growth and cell organization.

Some reports showed that PVA-base hydrogels could be used in all the above mentioned ways.

So, poly(vinyl alcohol) (PVA), physically crosslinked by repeated freeze-thawing cycles of polymer aqueous solutions, is widely employed to make hydrogels for biomedical applications. To increase the similarity between hydrogels and natural tissues and to obtain "polymeric hybrid tissues", 3T3 cells have been incorporated, from a mouse fibroblast cell line, into PVA hydrogels obtained by one freeze-thawing cycle using as a solvent complete culture medium [239]. Hydrogels were also made using eight freeze-thawing cycles from PVA solutions prepared using as a solvent either complete culture medium or water. Cell adhesion experiments were performed by seeding 3T3 and human umbilical vein endothelial cells (HUVEC) on to the hydrogel surface. The obtained results show that PVA is not cytotoxic. Although PVA hydrogel surface characteristics do not seem to favor the adhesion of substrate-dependent cells, encouraging results were obtained with the 3T3 cells incorporation. DMA analysis indicates that the networks prepared by eight freeze-thawing cycles possess a mechanical consistency comparable, even slightly better, than the ones prepared by only one freeze-thawing cycle and used for the cell incorporation studies.

Esmaiel Jabbari and Saeed Karbasi [240] noted that fibroblast cells seeded on N-vinyl pyrrolidone (NVP)-grafted PVA hydrogel, by using γ-radiation, had an extended oval morphology while those seeded on Acr.Ac.-grafted PVA had a rounded spherical morphology. These results support the use of NVP for grafting PVA to increase swelling and improve cell viability [240].

In order to achieve the firm fixation of the artificial cornea to host tissues, composites of collagen-immobilized poly(vinyl alcohol) hydrogel with hydroxyapatite were synthesized by a hydroxyapatite particles kneading method. The preparation method, characterization, and the results of corneal cell adhesion and proliferation on the composite material were studied. PVA-COL-HAp composites were successfully synthesized. A micro-porous structure of the PVA-COL-HAp could be introduced by hydrochloric acid treatment and the porosity could be controlled by the pH of the hydrochloric acid solution, the treatment time, and the crystallinity of the HAp particles. Chick embryonic keratocyto-like cells were well attached and proliferated on the PVA-COL-HAp composites. This material showed potential for keratoprosthesis application. Further study such as a long-term animal study is now required [241].

PVA/collagen substrate has been succesfull used also for osteoblasts grow [1].

Prosthesis, made by a composite body comprising polyvinyl alcohol hydrogel and ceramic or metallic porous body, has been proposed for a damaged bone, an artificial articular cartilage or an artificial intervertebral disc repairing. With this prosthesis, PVA hydrogel enhances lubrication and shock absorbing functions, and the porous body allows the ingrowth and ossification of the bone tissue of a living body therein to affinitively connect said hydrogel to the bones of the living body [242].

Nowadays the importance of knee meniscal function is recognized. The treatment for meniscus injury has been changing from resection to repair. However, depending on the type of injury, meniscectomy cannot be avoided. In consideration of the prognosis in such patients, artificial meniscus using polyvinyl alcohol-hydrogel (PVA-H) with high water content has been developed and performed an animal experiment as preliminary study. In the experiment using rabbits, the lateral meniscus was replaced with an artificial meniscus in one knee side and lateral meniscectomy was performed in another knee side of each rabbit. In the knees treated by artificial meniscus replacement, regressive changes were initially observed but did not progress after a certain period, and the articular cartilage state was good even after 1 year. In addition, neither wear nor breakage of PVA-H was observed. These results suggest that artificial meniscus using PVA-H with high water content compensates for meniscus function and is clinically applicable. However, for clinical application some problems such as fixation method, tolerance of PVA-H, remain to be solved [243,244].

To assess further the use of polyvinyl alcohol-hydrogel (PVA-H) artificial meniscus, some mechanical tests about PVA-H and animal experiment have been performed. In mechanical tests, it was found that a high water content PVA-H showed viscoelastic behavior similar to that of human meniscus. Moreover, the frictional coefficient of PVA-H against natural articular cartilage was also effective. In the animal experiment using rabbits, the lateral meniscus was replaced with an artificial meniscus in one knee side and lateral meniscectomy was performed in another knee side of each rabbit. In the results, the articular cartilage state of knee joint implanted PVA-H meniscus was good even after 2 years, while osteoarthrosis (OA) change progressed in meniscectomy knee joint. In addition, neither wear nor breakage of PVA-H was observed. These results proved that an artificial meniscus using a high water content PVA-H can compensate for meniscal function and might be clinically applicable [245,246].

The main disadvantage of hydrogels is their poor mechanical properties after swelling. In order to eliminate the disadvantage, hydrogels can be modified by physical blending [247,248] or/and chemical modification by grafting [249-251], crosslinking method [252-254] and semi-interpenetrating or interpenetrating polymer networks [255,256]. In order to overcome this difficulties, blends of PVA and chitosan have good mechanical properties and the applications of these blends have been reported [257,258] Chitosan (poly-β(1,4)-d-glucosamine), a cationic polysaccharide, is obtained by alkaline deacetylation of chitin, the principal exoskeletal component in crustaceans. As the combination of properties of chitosan such as water binding capacity, fat binding capacity, bioactivity, biodegradability, nontoxicity, biocompatibility, and antifungal activity, chitosan and its modified analogs have shown many applications in medicine, cosmetics, agriculture, biochemical separation systems, tissue engineering, biomaterials and drug controlled release systems [259-263]. Yang et al. [264] reported the preparation of PVA/chitosan blended membranes in various ratios and treated with formaldehyde. They were interested in studying the effect of chitosan content on the transport and equilibrium properties of membranes with of creatinine, uric acid and vitamin B12.

5-Fluorouracil (5-FU) is an antineoplastic agent that usually arrests tumor cells at the G1-S phase of the cell cycle and the choice in the treatment of carcinoma of colon or rectum; it is also used in the treatment of precancerous dermatoses, especially actinic keratosis for which is the treatment of choice if the lesions are multiple [265]. The cytotoxic anticancer drug often causes severe side effects because it does not act selectively on the target. In order to control the release rate of 5-FU, chitosan/ PVA blended hydrogel membranes can be used as the protective drug coatings. It was found that the water content and water vapour transmission rates on the blended hydrogel membrane increased with increasing chitosan content. In antibacterial assessment, the antibacterial activity of all chitosan/PVA blended hydrogel membranes is similar. The viable cell number of aurococcus on the various chitosan/PVA blended hydrogel membranes is about $(2.5 \pm 0.5) \times 10^7$ cells/mL. The authors show that permeability of solutes such as creatinine, 5-FU and vitamin B12 through chitosan/PVA blended hydrogel membranes increase linearly with chitosan content in the blended hydrogel membranes, whereas there is a sharp change of permeability of uric acid through the chitosan/ PVA blended hydrogel membrane when the chitosan content is changed from 60 to 80% in the blended hydrogel membrane.

3.3.4. Biomembranes in Artificial Organs

Hydrogel hybrid-type organs designed for implantation consists of living cells surrounded by suitable membranes. The living cells such as Langerhans islets, hepatoma (Hep G2), hepatocytes, etc, secreste specific compounds in response to the changes in body fluids. These systems work as a self-controlling bioreactor. The main point in design of an artificial organ is the choice of the suitable material and the preparation technique for membrane obtaining.

The main requirements for the membrane are:

- permeabilty against water, oxygen and nutrients;
- permeability for specific secretations of living cells;
- impermeability to components of the immune system;
- resistance to the biodegradability in the body conditions;
- non-adhesive for proteins (avoiding their deposition)

Artificial organs could be obtained by two main techniques:

1. microencapsulation of a small amount of living cells in microcapsules that will be injected into the organism;
2. design of a massive container with semipermeable membrane walls that contain a high number of living cells and that could be implanted in the peritoneal cavity acting as a substitute of the damaged organ [266].

The principle of a bioartificial pancreas design is presented in figure 11.

As it could be seen from the table 17, some methods of membranes obtaining could damage the living cells. The problems under study, related with artificial organs design are not only the obtaining of the suitable membrane, but also the possibilities of providing the suitable living conditions for the cells, avoiding their damage caused by the products of their methabolism.

A lot of reports dials with biomembranes application in artificial kidney and pancrease achieving and with haemodialysis problems [270-272] because hydrogels have the ability to swell in water and retain a significant fraction of water within its structure without dissolving and they have physical properties similar to those of human tissues and possesses excellent tissue compatibility.

Poly(vinyl alcohol) membranes satisfy the basic requirements for a bioartificial pancreas: good permeability for glucose, insulin and albumin but the passage of immunoglobulin G was completely prevented [273]. Furthermore, islets cultured in the PVA tubular membranes can perform their function of secreting insulin after 30 days in the static incubation study and rapidly releasing insulin through the membranes in response to changes in concentrations of glucose in the dynamic perifusion experiment [274]. Experimental data shown in Ref. 275 obtained from in vivo transplantation studies confirm that islets entrapped by the PVA tubular membrane chamber could change the glucose level in diabetic rats. When the m-2 [385] type of PVA chamber was implanted into streptozotocin induced diabetic rats, nonfasting blood glucose levels dropped from 500 ± 35 mg dL^{-1} to the lowest value (210 ± 22 mg dL^{-1}). Furthermore, the performance of the bioartificial pancreas can be enhanced by the increased numbers of implanted chambers. If three m-2 chambers were implanted, nonfasting blood glucose levels in the diabetic rats decreased to 130–160 mg dL^{-1} and such a low blood glucose value was maintained for 1 month. This indicates that implanting three m-2 chambers in the diabetic rats could provide improved permeability of insulin to normalize blood glucose levels and improved survival of islets from the immune system of the recipient. Therefore, this membrane provides adequate performance for secretory products in an application as a synthetic extracellular matrix for a bioartificial pancreas.

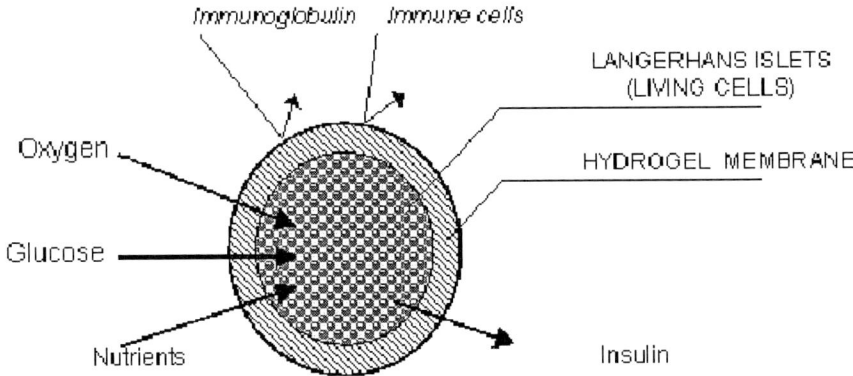

Figure 11. Scheme of a bioartificial pancreas surrounded by a hydrogel membrane [266].

Some methods used for artificial organs design are presented in table 17.

Table 17. Design of some artificial organs

Matrix	Crosslinker	Entrapped cells	Applications	Ref.
poly(allyl amine) and PVA as extracellular matrices	-	hepatocytes encapsulated in Ba-alginate capsules	-bio-artificial liver (BAL) that exhibit good metabolic functions such as albumin synthesis and ammonia removal	267
crosslinked alginate covered by a PVA membrane	-alginate is crosslinked by Ca^{2+} ions and PVA is crosslinked by GA	Langerhans islets	-bio-artificial pancreas -this procedure caused denaturation of cellular proteins	266
alginate matrix covered with PMMA membrane		Langerhans islets	-The shell has been deposit by interfacial precipitation -The cells are not damaged by this procedure and have a long term of survival	266
semipermeable membrane		pancreatic islet tissue	- The development of the bioartificial pancreas for treatment of human diabetes	268,269

3.4. CATALYSIS

Membrane reactors have found utility in a broad range of applications including biochemical, chemical, environmental, and petrochemical systems. The variety of membrane separation processes, the novel characteristics of membrane structures, and the geometrical advantages offered by the membrane modules have been employed to enhance and assist reaction schemes to attain higher performance levels compared to conventional approaches. In these, membranes in a reactor existing as membrane laminates or physically separated membranes with a fluid phase between them, can provide particular combinations for functions, such as separation of products from the reaction mixture, separation of a reactant from a mixed stream for introduction into the reactor, controlled addition of one reactant or two reactants, segregation of a catalyst (and cofactor) in a reactor, immobilization of a catalyst in (or on) a membrane. Membranes can act as both catalyst and reactor; membranes perform a wide variety of functions, often more than one function in a given context. Membranes in a reactor can be employed to introduce/separate/purify reactants and products, to provide the surface for reactions, to provide a structure for the reaction medium, or to retain specific catalysts [276].

Membranes can be used as a matrix for immobilization of a catalyst. Four basic types of catalysts are relevant: (a) enzymes and (b) whole cells for biocatalysis; (c) oxides and (d) metals for nonbiological synthesis. Biocatalysts will be considered first since their immobilization in (or on) the membrane was explored much earlier. Five techniques have been studied in varying degrees. They are (1) enzyme contained in the spongy fiber matrix; (2) enzyme immobilized on the membrane surface by gel polarization; (3) enzyme adsorbed on the membrane surface; (4) enzyme immobilized in the membrane pores by covalent bonding; (5) enzyme immobilized in the membrane during membrane formation by the phase inversion process of membrane making.

Membranes can also be used as a reactor where catalysts are used frequently. The membrane may physically segregate the catalyst in the reactor, or have the catalyst immobilized in the porous/microporous structure or on the membrane surface. The membrane having the catalyst immobilized in/on it acts almost in the same way as a catalyst particle in a reactor does, except that separation of the product(s) takes place, in addition, through the membrane to the permeate side. All such configurations involve the bulk flow of the reaction mixture along the reactor length while diffusion of the reactants/products takes place generally in a perpendicular direction to/from the porous/microporous catalyst.

PVA/chitosan blend membranes can be applied for the synthesis of monoglyceride, when used as a membrane enzyme reactor [277].

Lipases can catalyze hydrolysis of esters, synthesis of esters, trans-esterification, and synthesis of some polymers. They have been applied in many fields including the food industry, fine chemistry, and the pharmaceutical industry. The low stability of native lipases makes them unsuitable for industrial applications. In order to use them more economically and efficiently, their operational stability can be improved by immobilization. Numerous efforts have been focused on the preparation of lipases in immobilized forms involving a variety of both support materials and immobilization methods [278].

It was reported that PEGylated lipase entrapped in PVA cryogel could be conveniently used in organic solvent biocatalysis [279]. This method for enzyme immobilization is more convenient in comparison to other types of immobilization that take advantage of enzyme covalent linkage to insoluble matrix, since the chemical step which is time consuming and harmful to enzyme activity is avoided. The application of this catalytic system to the hydrolysis of acetoxycoumarins demonstrated the feasibility of proposed method in the hydrolysis products of pharmaceutical interest and to obtain regioselective enrichment of one of the two monodeacetylated derivatives.

Monoglyceride (MG) is one of the most important emulsifiers in food and pharmaceutical industries [280]. MG is industrially produced by transesterification of fats and oils at high temperature with alkaline catalyst. The synthesis of MG by hydrolysis or glycerolysis of triglyceride (TG) with immobilized lipase attracted attention recently, because it has mild reaction conditions and avoids formation of side products. Silica and celite are often used as immobilization carriers [281]. But the immobilized lipase particles are difficult to reuse due to adsorption of glycerol on this carriers [282]. PVA/chitosan composite membrane reactor can be used for enzymatic processing of fats and oils. The immobilized activity of lipase was 2.64 IU/cm2 with a recovery of 24%. The membrane reactor was used in a two-phase system reaction to synthesize monoglyceride (MG) by hydrolysis of palm oil, which was reused for at least nine batches with yield of 32–50%.

J. Xu et al. [283] have shown that immobilization of enzymes can be done using a specially designed composite membrane with a porous hydrophobic layer and a hydrophilic ultrafiltration layer. A polytetrafluoroethylene (PTFE) membrane with micrometer pores as an excellent hydrophobic support for immobilization was employed for the porous hydrophobic layer, and a biocompatible material of polyvinyl alcohol (PVA) which provided a favourable environment to retain the lipase activity was used to prepare the hydrophilic ultrafiltration layer. Enzyme molecules are adsorbed in pores of the hydrophobic layer and deposited on the interface between the hydrophilic layer and the hydrophobic layer by filtration. The PTFE layer supplied a large hydrophobic interface area to immobilize lipases which is beneficial for lipase activation [284]. The ultrafiltration PVA layer played a key role in controlling the enzyme loading and preventing enzymes from being dissolved into the aqueous phase. Furthermore, the mass-transfer resistance of water and water-solubility products through the hydrophilic membrane is lower than that through the hydrophobic dense cortical layer of an asymmetric membrane, which could reduce the negative effect of diffusion and product inhibition. A composite membrane with a porous PTFE layer and an ultrafiltration PVA layer demonstrated high efficiency in immobilizing *Candida rugosa* lipase. The immobilized enzyme membranes were used in a biphasic membrane reactor (BMR) for the hydrolysis of olive oil. The optimum enzyme loading per unit membrane area is 0.042 mg-protein cm^{-2}. In the

BMR, lipases on the surface of the membrane were removed by the flow of organic phase, but the flow of the organic phase does not decrease the activity of biocatalytic membranes. The lipase immobilized at the interface of the PTFE membrane and PVA layer are stable. The maximum reaction rate per unit of membrane area (9.25 $\mu mol\ h^{-1}\ cm^{-2}$) is be higher than the value reported in the literature [285] (2.18 $\mu mol\ h^{-1}\ cm^{-2}$) and [286] (1.77 $\mu mol\ h^{-1}\ cm^{-2}$). The immobilized lipase membrane in the BMR shows high activity for more than 30 h of reaction, with little change in the activity.

One of the extensively used synthetic polymers used as a support for immobilization of biocatalysts is polyacrylamide (PAAm) [287,288]. The major advantage is that it can be polymerized either chemically or by using radiation. Advantages of γ-ray polymerization against chemical polymerization is that the polymerization can be carried out even under frozen conditions thus allowing the matrix to be molded to any form such as beads or membranes [289-291]. However one of the major drawbacks of this polymer especially in a membranous form is its brittleness.

PVA has also been extensively used for immobilization of biocatalysts in a membranous form. As compared to PAAm, PVA is more hydrophilic and having adhesive property with better tensile strength in dry conditions. But it has high swelling index and dissolves readily in water when not cross-linked. PVA can be cross-linked using a variety of reagents including γ-rays.

PVA/acrylamide blend membranes prepared on cheese cloth support by γ-irradiation induced free radical polymerization can be used for urease entrapment. The enzyme urease is entrapped in the membrane during polymerization process and using glutaraldehyde as cross-linking agent. The main advantage of this blend to this process is that it can be reused a number of times without significant loss of urease activity [292].

But, glutaraldehyde (GA) is a well-known toxic reagent and its presence in the PVA matrix as residuals unremoved by washing procedures could damage the organism tissues. P.A. Ramires and E. Milella [293] proposed a technique of PVA/hyaluronic acid and PVA-gellan membranes crosslinking, by using GA in vapors state. They evaluated the harmful effects of GA residuals released from the membranes by the cytotoxicity and cytocompatibility in vitro tests, based on the cell culture method. The results showed that these materials have no toxic effects: they do not affect viability and proliferation nor exert damages on mithocondrial and lysosomal functions. The use of GA in vapor phase as crosslinking agent of natural and artificial polymer blends is demonstrated to be an effective way to avoid the presence of toxic residuals into materials [293].

3.5. PVA AND PVA DERIVATIVES-BASED MEMBRANES AS VAPORS AND GAS BARRIER

PVA films have high water-vapor permeability (water-vapor-transmission coefficient P_{H2O} is 270 g 0.1 mm/10 h m^2 cm-Hg) [294] that increases rapidly with relative humidity and with decrease in the hydrolysis degree.

PVA films have also a low gas permeability coefficient:

$P_{H2} = 6.6 \times 10^{-13}$ mL cm/(cm^2 s cm-Hg); $P_{O2} = 6.24 \times 10^{-17}$ mL cm/(cm^2 s cm-Hg); $P_{F2} < 10^{-13}$ mL cm/(cm^2 s cm-Hg); $P_{N2} = 10^{-11}$ mL cm/(cm^2 s cm-Hg); $P_{CO2} = 10^{-13}$ mL cm/(cm^2 s cm-Hg) [294].

The gas permeability increases with increase the relative humidity, with decrease hydrolysis degree of PVA, with increase temperature, and tends to decrease sharply as the degree of crystallinity increases. The decrease in crystallinity and the decrease in T_g would be expected to increase the gas permeability.

Because of its low gas permeability, PVA has excellent flavour-retaining properties [18].

Physical crosslinked PVA cryogel is considered to have a good permeability for oxygen that is a desirable property for biomaterials [1].

Sometimes, in different systems, the oxygen presence is undesirable because of its reactivity and tendency to oxidize the contact materials that leads to corrosion of metallic materials or depreciation of food quality. Also oxygen could inhibit different chemical reactions or could interfere in different analysis (RES, polaroghaphy, etc.).

Due to these practical aims, membranes with low oxygen permeability have been developed. Some of them are PVA, PVA blends or their derivative membranes, due to the PVA excellent oxygen barrier properties [18].

One of the PVA derivatives extensively used in this field is ethylene-vinyl alcohol copolymer (EVOH). Their blends with different polyolefins are also effective as oxigen barrier materials.

Blend film oxygen permeability is influenced by the film composition and morphology. Generally, a heterogeneous structure, containing orientated fibrils and lamellae of EVOH evidences lower oxygen permeability than that emphasized by a more homogeneous morphology with finer dispersed particles (table 13 [295]).

The O_2 permeability of the blends obtained in the batch mixer decreases (from 59.1 to 47.7 mm cm^3/m^2 day^{-1} atm^{-1} for EVOH / PPlv and from 53.6 to 43.3

mm cm³/m² day⁻¹ atm⁻¹ for EVOH/ PPhv) with the increasing of EVOH content from 12.5 to 25 vol%.

The permeability for the extruded films was lower that those of the pressed films (22.1 mm⁻¹ cm³/m² day⁻¹ atm⁻¹ for 10 vol. % EVOH).

Dry EVOH/PP-g-maleic mldehyde (MAH) blend, obtained by moulded injection also evidenced good barrier properties for toluene. This property is improved by increasing EVOH concentration which determines both size and deformation of the minor phase increase, indicating that the laminar structure becomes more pronounced [296]. But even the laminar structure is maximized a moulded injection sample is not likely to reach a permeability as low as expected for a multilayered system [296].

Oxygen and toluene permeability of the blend film decreases as a nonlinear function with increasing the EVOH content, as it may be seen in table 19.

Table 18. Influence of the films "composition and of the blend morphology on the films" permeability to O_2 [295]

Morphology	EVOH / (%)	HDPE / (%)	PPLv / (%)	PPhv / (%)	PP-g-MAH (%)	P_{O_2} / / cm³ mm/ (m² day⁻¹ atm⁻¹)	Remarks
Fine dispersed particles	12-25	-	X	-	-	58-80	Permeability independent of EVOH conc.
	12-25	-	-	X	-	65-230	Permeability independent of EVOH conc.
	>25	-	-	X	-	>65	-interface between large particles of EVOH and PP can run from one side to the other of the film, created voids increasing the permeability.
	<12	-	-	X	-	<65	-smaller EVOH particles.
	<12	-	X	-	-	<70	-lover level of voids
Fibrils and lamellae	10	-	10	80	-	22-25	-draw ratio=3.4
	10	-	90	-	-	22.1	-draw ratio=2.8
	10	-	80	-	10	23.1-25.5	-draw ratio=2.8-8.7
	20	-	80	-	-	12.4	-draw ratio=3.4
	20	20	60	-	-	51.6	-draw ratio=3.4 -poor adhesion between PP and HDPE interfaces.
	20	-	60	-	20	9.5	-draw ratio=3.2 -lamellae coexists with fibrils

HDPE= high density polyethylene; PPLv= polypropylene with low viscosity; PPhv= polypropylene with high viscosity; PP-g-MAH= polypropylene graft maleic aldehyde.

Table 19. Oxigen and toluene permeability of melt-blended EVOH-Nylon 6(L), measured at 30 °C [297]

EVOH/Nylon 6(L)	P(oxygen)×10^{13} cm^3 cm s^{-1} cm^{-2} cm-Hg^{-1}	P (toluene) g mm m^{-2} 24h^{-1}
100/0	0.31	0.08
75/25	0.57	0.10
50/50	1.63	0.14
25/75	4.87	0.23
0/100	13.79	0.29

The EVOH-COOH compatibilizer use determines the increasing of the blend film oxygen permeability that becomes two orders of magnitude higher than that of a coextruded film with the same percentage of EVOH [298]. That means that EVOH-COOH acts that an interfacial agent.

Excessive emission of CO_2 has caused the most dramatic increase in global atmospheric temperature. So many countries' governments and researchers pay much attention to how to predict, control and reduce the amount of CO_2, emission. Compared to the absorption and adsorption techniques, membrane processes can be operated continuously and require less energy for the separation or purification. However, commercial polymer membranes cannot achieve both enough high selectivity and permeability to meet these needs. L. Xu et al. [299] have shown that polyvinylidene fluoride (PVDF)-PVA hydrogel membranes contained sodium carbonate solution and immobilized carbonic anhydrase can be used for removing CO_2 in air.

Hydrogel membranes are particularly attractive because of high permeability and separation factor [300], and good stability for CO_2/N_2 separation [299]. PVDF hollow fiber membrane modified by alkali was coated by PVA hydrogel on its surface and PVDF-PVA hydrogel membranes show better hydrophilic performance. For carbonate hydrogel (sodium carbonate concentration of 3.7 %) membrane, CO_2, concentration from 1.3 % to 0.6 % in feed gas could be decreased to 0.9-0.4 % at the outlet at 25 °C.

PVA/CELL blend could be also used to obtain the membranes with a low permeability for CO_2 [301].

In the last years, the replacement of gasoline with other new fuels became a priority because of unavoidable depletion of natural petroleum sources.

The methanol/gasoline fuel has been proved to be one of the best replacements for gasoline because of its low cost, high efficiency and low air pollution. Because of the corrosive character of methanol, the metal vials for storing fuels have to be changed. Polyolefins such as high density polyethylene (HDPE) have been considered as a potential material for methanol/gasoline fuel storrage because of their low cost, lightweight, easy design and processing, recyclability, safety, high chemical resistance to corrosion, and flexibility. An important draw back of the HDPE use for this purpose is its poor permeation resistance to hydrocarbon solvents, such as gasoline. Escaping of the gasoline vapor into the atmosphere could determine serious environmental pollution.

Many efforts have been directed to finding methods that could reduce the HDPE permeability. One of them is HDPE blending with polymers such as polyamide (PA) or PVA with low permeability for hydrocarbons. To overcome the incompatibility between the non-polar and polar polymers, a compatibilizer has to be present in these blends.

The high permeation resistance of the polymeric blend against the hydrocarbons depends on the blend composition but also on the obtaining technique. So, multi-layer co-extrusion of PE, compatibilizer precursor (CP) and PA, laminar blend blow molding of PE, CP and PA blends, laminar blend blow molding of PE and modified PA (MPA) have been applied for low permeation materials obtaining. Because PA and CP did not sufficiently increase the HDPE permeation resistance, PVA has been introduced in the blend because of its recognized high barrier qualities.

Good methanol/gasoline fuel permeation resistance together with clearly defined MPAPVA and MPA laminar structures were found in containers blow-molded from PE/PMPAPVA and PE/MPA blends, respectively, with an optimum CP of about 20 wt% [302].

Chapter 4

CONCLUSION

In the recent years, many researchers have devoted attention to the development of membrane science and technology. Different important types of membranes, such as these for: nanofiltration, ultrafiltration, microfiltration, separation of gases and inorganic membranes, facilitated or liquid membranes, catalytic and conducting membranes, and their applications and processes, such as wastewater purification and bio-processing have been developed [303]. In fact, almost 40 % of the sales from membrane production market are for purifying wastewaters.

Poly(vinyl alcohol) (PVA) has been characterized on many levels and examined for numerous applications. It is a polymer of great interest because of its relatively simple structure, easy processing, and potential use in biomedical and pharmaceutical fields. The possibilities to control the PVA's biodegradability make from PVA a friendly polymer. Also the large possibilities to modify and control the PVA properties starting from synthesis process (such as molar mass, hydrolysis degree, OH groups repartition on the polymeric chain, tacticity) and also from its capability to react with a lot of reagents leading to polymer analogous compounds or to be cross-linked by chemical or physical ways make from PVA a versatile product. Also its capacity to be blend with other polymers or to be copolymerized with different co-monomers or to be doped with organic or inorganic compounds or to encapsulate drugs or enzymes enlarges the possibilities of PVA-based materials use. PVA capability of film formation, its mechanical resistance, high optical properties and the capacity of its hydrogels to swell in water and the hydrogels high sensitivity to the environmental alterations could characterize PVA as an intelligent material with special properties that could be tailored in function of the use interest.

Hydrogel membranes fulfill many of the important conditions for most of above-mentioned application fields. Therefore, we have focused our paper on the applications of PVA-based membranes in areas such as for separation membranar processes, fuel cells, sensors, biochemical/medical applications, catalyst or PVA derivatives membranes as gas and vapor barriers.

However, PVA is also extensively used in other different forms, such as gel matrices, micro and nano spheres, aerosols, aqueous solution, films, powder etc. Although some of the different types of PVA gels have been referred in chapter 3.3, it still remains much more to say. This clearly proves that PVA is an old, yet new polymer or, in other words, an old polymer with a promising future, due to its capacity to respond to all the actual society priorities: clean technologies, non-toxicity, biocompatibility, biodegradability, intelligent materials.

REFERENCES

[1] C.H. Hassan, N.A. Peppas. *Adv. Polym. Sci.* 2000;153:37.
[2] N.A. Peppas, "Hydrogels in medicine and pharmacy", in *"Polymers"*, N.A. Peppas (ed.), vol. 2, CRC, Boca Raton, FL, 1987.
[3] Danno. J. Phys. *Soc. Jpn.* 1958; 13:722.
[4] N.A. Peppas, E.W. Merrill. *J. Appl. Polym. Sci.* 1976;20:1457.
[5] J.C. Bray, E.W. Merrill. *J. Appl. Polym. Sci.* 1973;17:3781.
[6] N.A. Peppas, E.W. Merrill. *Techn. Chron.* 1974;43:559.
[7] S.R. Stauffer, N.A. *Peppas. Polymer.* 1992;33:3932.
[8] C.H. Hassan, N.A. Peppas. *Macromolecules.* 2000;33:2472.
[9] N.A. Peppas, S.R. Stauffer. *J. Controlled Release.* 1991;16:305.
[10] V.I. Lozinsky. *Russian Chem. Rev.* 1998;67:573.
[11] C.C. DeMerlis, D.R. Schoneker. *Food Chemical Toxicology.* 2003;41:319.
[12] D. Reddy, C.E. Reineke. *AIChE Symp. Ser.* 1998;84:84.
[13] E. Bengtsson, G. Traardh, B. Hallstrom. *J. Food Eng.* 1993;19:399.
[14] L. Enneking, W. Stephan, A. Heintz. *Ber. Bunsenges. Phys. Chem.* 1993;97:912.
[15] U. Sander, P. Soukup. *J. Membr. Sci.* 1998;36:463.
[16] Smitha, D. Suhanya, S. Sridhar, M. Ramakrishna. *J. Membr. Sci.* 2004;241:1.
[17] J.W. Rhim, M.Y. Sohn, K.H. Lee. *J. Appl. Polym. Sci.* 1994;52:1217.
[18] S. Patachia, "Blends based on poly(vinyl alcohol) and the products based on this polymer", in *"Handbook of Polymer blends and composites"*, C. Vasile and A.K. Kulshreshtha (eds.), Chap. 8, RAPRA Technology LTD., England, Chap.8. 2003. p. 288-365.
[19] I.J. Ball, S.C. Huang, R.A. Wolf, J.Y. Shimano, R.B. Kaner. *J. Membr. Sci.* 2000;174:161.

[20] S.C. Huang, I.J. Ball, R.B. Kaner. *Macromolecules.* 1998;31:5456.
[21] E.M. Genies, A. Boyle, M. Lapkowski, C. *Tsintavis. Synth. Met.* 1990;36:139.
[22] F. Lux. *Polymer.* 1994;35:2915.
[23] H.G. Neoh, E.T. Kang, K.L. Tan. *Polym. Degrad. Stab.* 1993;40:357.
[24] E.T. Kang, K.G. Neoh, K.L. Tan. *Prog. Polym. Sci.* 1998;23:277.
[25] Alix, V. Lemoine, M. Nechtschein, J.P. Travers, C. Mendaro. *Synth. Met.* 1989;29:457.
[26] M.J. Liu, K. Tzon, R.V. Grefory. *Synth. Met.* 1994;63:67.
[27] B.V.K. Naidu, M. Sairam, K.V.S.N. Raju, T.M. Aminabhavi. *J. Membr. Sci.* 2005;260:142.
[28] M. Sairam, M.B. Patil, R.S. Veerapur, S.A. Patil, T.M. Aminabhavi. *J. Membr. Sci.* 2006;281:95.
[29] X. Chen. *J. Mater. Sci. Lett.* 2002;21:1637.
[30] D.J. Upadhyay, N.V. Bhat. *J. Membr. Sci.* 2004;239:255.
[31] J. Upadhyay, N. V. Bhat. *J. Membr. Sci.* 2004;255;181.
[32] S.S. Kulkarni, A.A. Kittur, M.I. Aralaguppi, M.Y. Kariduraganavar. *J. Appl. Polym.* Sci. 2004;94:1304.
[33] V.V. Namboodiri, R.Ponangi, L.M. Vane. *Eur. Polym. Jnl.* 2006;42:3390.
[34] R. Psaume, Y. Aurell, J.C. Mora, J.L. Bersillon. *J. Membr. Sci.* 1988;36:373.
[35] T. Uragami, K. Okazaki, H. Matsugi, T. Miyata. *Macromolecules.* 2002;35:9156.
[36] S.K. Mallapragada, N.A. Peppas, *J. Polym Sci., Part B:Polym. Phys.* 1996;34:1339.
[37] M. Rafik, A. Mas, M.-F. Guimon, C. Guimon, A. Elharfi1, F. *Schu. Polym. Int.* 2003;52:1222.
[38] K. Benzekri, A. Essamri, N. Toreis, A. Souissi, T. Maarouf, A. Mas. *Eur. Polym. Jnl.* 2001;37:1607.
[39] W.P. Chang, W.T. Whang. *Polymer.* 1996;37:3493.
[40] H.C. Park, R.M. Meertens, M.H.V. Mulder, C.A. Smolders. *J. Membrane Sci.* 1994;90:265.
[41] J.W. Rhim, M.Y. Sohn, K.H. Lee. *J. Appl. Poym. Sci.* 1994;52:1217.
[42] J.W. Rhim, H.K. Kim, K.H. Lee. *J. Appl. Polym. Sci.* 1996;61:1767.
[43] K.H. Lee, H. Kim, J.W. Rhim. *J. Appl. Poym. Sci.* 1995;58:1707.
[44] M. Metayer, C.O. Mbareck. *Reactive Functional Polym.* 1997;33:311.
[45] W. Herrera-Kao, M. Aguilar-Vega. *Polym. Bulletin.* 1999;42:449.
[46] E. Ruckenstein, Y. Sun. *J. Appl. Polym. Sci.* 1996;61:1949.
[47] E. Ruckenstein, L. Liang. *J. Appl. Polym. Sci.* 1996;62:973.

[48] J.G. Byun, Y.M. Lee, C.S. Cho. *J. Appl. Polym. Sci.* 1996;61:697.
[49] M. Suzuki, T. Tateishi, M. Matsuzawa, M. Saito. *Macromol. Symp.* 1996;109:55.
[50] T. Hirai, T. Okinaka, Y. Amemiya, K. Kobayashi, M. Hirai, S. Hayashi. *Angew. Makromol. Chem.* 1996;240:213.
[51] Y.M. Lee, S.H. Kim, S.S.Cho. *J. Appl. Polym. Sci.* 1996;62:301.
[52] J.W. Rhim, S.W.Yoon, S.W. Kim, K.H. Lee. *J. Appl. Polym. Sci.* 1997;63:521.
[53] A.S. Hickey, N.A. Peppas. *Polymer.* 1997;38:5931.
[54] A.V. Volkov, I.V. Karachevtsev, M.A. Moskvina, A.V. Rebrov, A.L.Volinskii, N.F. Bakeev, *J. Inorganic Organometallic Polym.* 1995;5:295.
[55] C. Vauclair, H. Tarjus, P. Schatzel. *J. Membrane Sci.* 1997;125:293.
[56] H.S. Shin, S.Y. Kim, Y.M. Lee. *J. Appl. Polym. Sci.* 1997;65:685.
[57] C.Vasile, E.M. Calugaru, S.F. Bodonea, *J. Polymer Sci.: Polymer Chemistry. Ed.* 1981;19:879.
[58] J.J. Shieh, R.Y.M. Huang. *J. Membr. Sci.* 1998;148:243.
[59] X.P. Wang, Z.Q. Shan, F.Y. Zhang, Y.F. Zhang. *J. Appl. Polym. Sci.* 1999;73:1145.
[60] Chanachai, R. Jiraratananon, D. Uttapap, G.Y. Moon, W.A. Anderson, R.Y.M. Huang. *J. Membr. Sci.* 2000;166:271.
[61] Y.M. Lee, S.Y. Nam, D.J. Woo. *J. Membr. Sci.* 1997;133:103.
[62] T. Uragami, T. Matsuda, H. Okuno, T. Miyata. *J. Membr. Sci.* 1994;88:243.
[63] T. Uragami, K. Takigawa. *Polymer.* 1990;31:668.
[64] B.-B. Lia, Z.-L. Xua, F.A. Qusaya, *R. Lic. Desalination.* 2006;193:171.
[65] C.K. Yeom, K.H. Lee. *J. Appl. Polym. Sci.* 1998;67:209.
[66] M.D. Kurkuri, U.S. Toti, T.M. Aminabhavi. *J. Appl. Polym. Sci.* 2002;86:3642.
[67] J. Joncceon, K.H. Lee. *J. Appl. Polym. Sci.* 1996;61:389.
[68] Y.Q. Dong, L. Zhang, J.N. Shen, M.Y. Song, H.L. Chen. *Desalination,* 2006;193:202.
[69] M.Y.Kariduraganavar, S.S. Kulkarni, A.A. Kittur. *J. Membrane Sci.* 2005;246:83.
[70] N. Alghezawi, O. Sanh, L. Aras, G. Asman. *Chemical Eng. Processing,* 2005;44:51.
[71] L. Zang, P.Yu, Y. Luo, *Sep. Purif. Technol.* 2006 (in press).
[72] S.M. Ahn, J.W.Ha, J.H. Kim, Y.T. Lee, S.B. Lee. *J. Membrane Sci.* 2005;247:51.
[73] D. Graiver, S.H. Hyon, Y. Ikada. *J. Appl. Polym. Sci.* 1995;57:1299.

[74] N.D. Hilmioglu, S. Tulbentci, *Desalination.* 2004;160:263.
[75] J.W. Rhim, Y.K. Kim. *J. Appl. Polym. Sci.* 2000;75:1699.
[76] J.P.G. Villaluenga, A. Tabe-Mohammadi. *J. Membr. Sci.* 2000;169:159.
[77] Yamasaki, T. Shinbo, K. Mizoguchi. *J. Appl. Polym. Sci.* 1997;64:1061.
[78] F. Peng, L. Lu, C. Hu, H. Wu, Z. Jiang. *J. Membr. Sci.* 2005;259:65.
[79] F. Peng, Z. Jiang, C. Hu, Y. Wang, L. Lu, H. Wu. *Desalination.* 2006; 193:182.
[80] F. Peng, L. Lu, H. Sun, Y. Wang, J. Liu, Z. Jiang. *Chem. Mater.* 2005;17:6790.
[81] L. Lu, H. Sun, F. Peng, Z. Jiang. *J. Membr. Sci.* 2006;281:245.
[82] J.N. Shen, L.G.Wu, H.L Chen, C.J. Gao. *Sep. Purif. Technol.* 2005;45:103.
[83] T. Miyata, T. Iwamoto, T. Uragami. *Macromol. Chem. Physics.* 1996;197:2909.
[84] T. Miyata, T. Iwamoto, T. Uragami. *J. Appl. Poym. Sci.* 1994;51:2007.
[85] K. Sreenivasan, *J. Appl. Polym. Sci.* 1997;65:1829.
[86] N.S. Rathore, J.V. Sonawane, A. Kumar, A.K. Venugopalan, R.K. Singh, D.D. Bajpai, J.P. Shukla. *J. Membr. Sci.* 2002;189:119.
[87] S. Schlosser, R. Kertész, J. Martak. *Sep. Purif. Technol.* 2005;41:237.
[88] S. Touil, S. Tingry, S. Bouchtallaa, A. Deratani. *Desalination.* 2006;193:291.
[89] S. Touil, S. Tingry, J. Palmeri, S. Bouchtalla, A. Deratani. *Polymer.* 2005;46:9615.
[90] M. Di Luccio, B.D. Smith, T. Kida, T.L.M. Alves, C.P. Borges. *Desalination.* 2002;148:213.
[91] S. Patachia, L. Isac, M. Rinja. Environm. *Eng. Management J.* 2004;3:661.
[92] S. Patachia, M. Rinja, L. Isac. *Rom. Journ. Phys.* 2006;51:253.
[93] M. Rinja, S. Patachia. Synthesis, characterization and applications of PVA hydrogels. *Pollack Periodica.* 2007. In press.
[94] S. Varga, S. Patachia, R. Ion. The application of poly(vinyl alcohol) based hydrogels for the decontamination of porphyrins-containing medical waste waters. Endvedu-2007 International Conference, Brasov, Romania, Book of Abstracts; *Bulletin of the "Transilvania"* University of Brasov, 2007. In press.
[95] J. Szejtli. Cyclodextrins Technology. *Kluwer Acad. Publ., Dordrecht,* 1988.
[96] S. Li, W.C. Purdy. *Chem. Rev.* 1992;92:1457.
[97] K. Uekama, F. Hirayama, T. Irie. *Chem Rev.* 1998;98:2045.
[98] P.C. Manor, W. Saenger. *J. Am. Chem. Soc.* 1974;96:3630.
[99] M. Nilsson, C. Cabaleiro-Lago, A.J.M. Valente, O. Söderman. *Langmuir.* 2006;22:8663.

[100] J. M. Valente, M. Nilsson, O. Söderman. *J. Colloid Interf. Sci.* 2005;281:218.
[101] A.C.F. Ribeiro, M. Esteso, V.M.M. Lobo, A.J.M. Valente, S.M.N. Simões, A.J.F.N. Sobral, M.L. Ramos, H.D. Burrows, A.M. Amado, A.M. Amorim da Costa, *A.M. J. Carbohydrate Chem.* 2006;25:173.
[102] Z. Juvancz, J. Szejtli. *Trends Anal. Chem.* 2002;21:379–88.
[103] M. Kim, J.D. Way, R.M. Baldwin. *Korean J. Chem. Eng.* 2002;19:876.
[104] H.L. Chen, L.G. Wu, J. Tan, C.L. Zhu. *Chem. Eng. J.* 2000;78:159.
[105] S.P. Kusumocahyo, T. Kanamori, K. Sumaru, T. Iwatsubo, T. Shinbo. *J. Membr. Sci.* 2004;231:127.
[106] Yamasaki, T. Iwatsubo, T. Masuoka, K. Mizoguchi. *J. Membr. Sci.* 1994;89:111.
[107] S. Patachia, M. Voinea. Biological materials as solution for water depollution. Buletinul Institutului Politehnic din Iasi, Tomul LI (LV), Fasc. 4, 2005, *Sectia Stiinta si Ingineria Materialelor,* p. 199-203.
[108] S. Patachia, M. Voinea. Bioaccumulation as a technique of cations separation from aqueous solution. Proceedings, EnvEdu-2005, *Trends in environmental Education,* Ed. Univ. 2005. p.54.
[109] A.S. Jonsson, G. Tragardh. *Desalination.* 1990;77:135–179.
[110] M.D. Afonso, R. Borquez. *Desalination.* 2002;142:29.
[111] Akthakul, W.F. McDonald, A.M. Mayes. *J. Membr. Sci.* 2002;208:147.
[112] S. Nakao. *J. Membr. Sci.* 1994;96:131.
[113] A.D. Marshall, P.A. Munro, G. Tragardh. *Desalination.* 1993;91:65.
[114] R.H. Li, T.A. Barbari. *J. Membrane Sci.* 1995;105:71.
[115] J. She, X. Shen, *Desalination.* 1987;62:395.
[116] M.G. Katz, T. Wydeven. *J. Appl. Poly. Sci.* 1982;27:79.
[117] L. Na, Z. Liu. *J. Membr. Sci.* 2000;169:17.
[118] X. Wang, D. Fang, K. Yoon, B.S. Hsiao, B. Chu. *J. Membr. Sci.* 2006;278:261.
[119] W.S. Dai, T.A. Barbari. *J. Membr. Sci.* 1999;156:67.
[120] B. Ding, H. Kim, S. Lee, C. Shao, D. Lee, S. Park, G. Kwag, K. Choi. *J. Polym. Sci.: Part B: Polym. Phys.* 2002;40:1261.
[121] S.P. Nunes, M.L. Sforca, K.-V. Peinemann. *J. Membr. Sci.* 1995;106:49.
[122] Pinnau, B.D. Freeman. *Polym. Mater. Sci. Eng.* 2002;86:108.
[123] Pinnau, B.D. Freeman. Advanced Materials for Membrane Separations. Oxford University Press, Oxford, 2004.
[124] Y. Zhang., H. Li, H. Li, R. Li, C. Xiao. *Desalination.* 2006;192:214.
[125] Y. Tsai, S. Li, *J. Chen. Langmuir.* 2005;21:3653.
[126] R. Mukundan, E. Brosha, F. Garzon. *Solid State Ionics.* 2004;175:497.

[127] Yamauchi, K. Togami, A.M. Chaudry, A.M. El Sayed. *J. Membr. Sci.* 2005;249:119.
[128] Fuel Cell Handbook, 6th ed., *B/T Books, Orinda, CA,* 2002.
[129] Aramata, I. Toyoshima, M. Enyo. *Electrochim. Acta.* 1992;37:1317.
[130] J. Wang, S. Wasmus, R.F. Savinell. *J. Electrochem. Soc.* 1995;142:4218.
[131] M.K. Ravikumar, A.K. Shukla. *J. Electrochem. Soc.* 1996;143:2601.
[132] T. Schultz, S. Zhou, K. Sundmacher. *Chem. Eng. Technol.* 2001;24:1223.
[133] X. Ren, T.A. Zawadzinski, F. Uribe, H. Dai, S. Gottesfeld. *Electrochem. Soc., Proc.* 1995;95:284.
[134] R. Nolte, K. Ledjeff, R.M. Baue, R. Mulhaupt. *J. Membr. Sci.* 1993 ;83:211.
[135] J. Kerres, A. Ulrich, F. Meier, T. Haring. *Solid State Ionics.* 1999;125:243.
[136] T. Kobayashi, M. Rikukawa, K. Sanui, N. Ogata. *Solid State Ionics.* 1998;106:219.
[137] R.W. Kopitzke, C.A. Linkours, H.R. Anderson, G.L. Nelson. *J. Electrochem. Soc.* 2000;147:1677.
[138] M. Kawahara, M. Rikukawa, K. Sanui, N. Ogata. *Solid State Ionics.* 2000;136:1193.
[139] J.P. Randin. *J. Electrochem. Soc.* 1982;129:1215.
[140] W. Charles, W. Walker Jr. *J. Power Sources.* 2002;110:144.
[141] J.A. Kerres. *J. Membr. Sci.* 2001;185:3.
[142] M.-S. Kang, Y.-J. Choi, S.-H. Moon. *J. Membr. Sci.* 2002;207:157.
[143] S.D. Mikhailenko, K. Wang, S. Kaliaguine, P. Xing, G.P. Robertson, M.D. Guiver. *J. Membr. Sci.* 2004;233:93.
[144] J. Qiao, T. Hamaya, T. Okada. *Chem. Mater.* 2005;17:2413-2421.
[145] J. Qiao, T. Hamaya, T. Okada. *Polymer.* 2005;46:10809.
[146] B.R. Einsla, Y.S. Kime, M.A. Hickner, Y.T. Hong, M.L. Hill, B.S. Pivovar, J.E. McGrath. *J. Membr. Sci.* 2005;255:141.
[147] C. Chanthad, J. Wootthikanokkhan. *J. Appl. Polym. Sci.* 2006;101:1931.
[148] N.W. DeLuca, Y.A. Elabd. *J. Memb. Sci.* 2006;282:217.
[149] D. Kuckling, A. Richter, K.-F. Arndt. *Macromol. Mater. Eng.* 2003;288:144.
[150] Richter, D. Kuckling, K.-F. Arndt, T. Gehring, S. Howitz. *J. Microelectromech. Syst.* 2003;12:748.
[151] K.-F. Arndt, A. Richter, S. Ludwig, J. Zimmermann, J. Kressler, D. Kuckling, H.-J. Adler. *Acta Polym.* 1999;50:383.
[152] D.J. Beebe, J.S. Moore, J.M. Bauer, Q. Yu, R.H. Liu, C. Devadoss, B.-H. Jo. *Nature.* 2000;404:588.
[153] R. Bashir, J.Z. Hilt, O. Elibol, A. Gupta, N.A. Peppas. *Appl. Phys. Lett.* 2002;81:3091.

[154] K.-F. Arndt, D. Kuckling, A. Richter. *Polym. Adv. Technol.* 2000;11:496.
[155] K.-F. Arndt, T. Schmidt, H. Menge. *Macromol. Symp.* 2001;164:313.
[156] X. Liu, X. Zhang, J. Cong, J. Xu, K. Chen. *Sens. Actuat.* B 2003 ;96:468.
[157] A.J. Marshall, J. Blyth, C.A.B. Davidson, C.R. Lowe. *Anal. Chem.* 2003;75:4423.
[158] Richter, A. Bund, M. Keller, K.-F. Arndt. *Sens. Actuat.* B 2004;99:579.
[159] M. Guenther, G. Suchaneck, J. Sorber, G. Gerlach, K.-F. Arndt, A. Richter. *Fine Mech. Opt. (Olomouc)* 2003;48:320.
[160] S.K. De, N.R. Aluru, B. Johnson, W.C. Crone, D.J. Beebe, J. Moore. *J. Microelectromech. Syst.* 2002;11:544.
[161] G. Gerlach, M. Guenther, J. Sorber, G. Suchaneck, K.-F. Arndt, A. Richt. *Sens. Actuat. B.* 2005;111–112:555.
[162] S.Patachia, M. Rinja. Study of the PVA hydrogel behaviour in boric acid solution. *Advances in Micro and Nanoengeneering,* Series Micro and Nanoengineering 6, Ed. Academiei Romane, 2004. pp. 140-146.
[163] Wolfbeis. *Anal. Chem.* 2002;74:2663.
[164] J. Lin. *Trends Anal. Chem.* 2000;19:541.
[165] Lobnik, I. Oehme, I. Murkovic, O.S. Wolfbeis. *Anal. Chim. Acta.* 1998;367:159.
[166] G.J. Mohr, O.S. Wolfbeis. *Anal. Chim. Acta.* 1994;292:41.
[167] Y. Kostov, S. Tzaonkov, L. Yotova, M. Krysteva. *Anal. Chim. Acta.* 1993;280:15.
[168] Z.H. Liu, F.L. Luo, T.L. Chen. *Anal Chim Acta.* 2004;510:189.
[169] Z.H. Liu, F.L. Luo, T.L. Chen. *Anal. Chim. Acta.* 2004;519:147.
[170] Z. Liu, J. Liu, T. Chen. *Sens. Actuat.* B 2005;107:311.
[171] S.S. Rosatto, P.T. Sotomayor, L.T. Kubota, Y. Gushikem. *Electrochim. Acta.* 2002;47:4451.
[172] F. Gelman, J. Blum, D. Avnir. *J. Am. Chem. Soc.* 2002;124:14460.
[173] J. Wang, P.V.A. Pamidi, D. Su Park., *Anal. Chem.* 1996;68:2705.
[174] E.J. Cho, Z. Tao, E.C. Tehan, F.V. Bright. *Anal. Chem.* 2002;74:6177.
[175] Kumar, R. Malhotra, B.D. Malhotra, S.K. Grover. *Anal. Chim. Acta.* 2000;414:43.
[176] Gill, *Chem. Mater.* 13 (2001) 3404.
[177] W. Jin, J.D. Brennan. *Anal. Chim. Acta.* 2002;461:1.
[178] S. Cosnier, A. Senillou, M. Gratzel, P. Comte, N. Vlachopoulos, N.-J. Renault, C. Martelet. *J. Electroanal. Chem.* 1999;469:176.
[179] B. Prieto-Simon, G. Armatas, Ph.J. Pomonis, C.G. Nanos, M.I. Prodromidis. *Chem. Mater.* 2004;16:1026.
[180] Silvana, L. Barthelmebs, J.-L. Marty. *Anal. Chim. Acta.* 2002;464:171.

[181] K.S. Chang, W.L. Hsu, H.Y. Chen, C.K. Chang, C.Y. Chen. *Anal. Chim. Acta.* 2003;481:199.
[182] F. Mizutani, T. Sawaguchi, Y. Sato, S. Yabuki, S. Iijima. *Anal. Chem.* 2001;73:5738.
[183] C.G. Tsiafoulis, M.I. Prodromidis, M.I. Karayannis. *Biosen. Bioelectron.* 2004;20:620.
[184] W. Sant, M.L. Pourciel-Gouzy, J. Launay, T. Do Conto, R. Colin, A. Martinez, P. Temple-Boyer. *Sens. Actuat. B* 2004;103:260.
[185] B.K. Sohn, B.W. Cho, C. S. Kim, D. H. Kwon. *Sens. Actuat. B* 1997;41:7.
[186] N. Jaffrezic-Renault, A. Senillou, C. Martelet, K. Wan, J. M. Chovelon, *Sens. Actuat. B* 1999;59:154.
[187] R. Rouillon, M. Tocabens, J. L. Marty. *Anal. Lett.* 1994;27:2239.
[188] J. Perdomo, H. Hinker, H. Sundermeier, W. Seifert, O. Martinez, M. Knoll. *Biosens. Bioelectron.* 2000;14:515.
[189] F. Mizutani, Y. Sato, T. Sawaguchi, S. Yabuki, S. Iijima. *Sens. Actuat. B* 1998;52:23.
[190] D. T. V. Anh, W. Olthvis, P. Bergveld. *Sens. Actuat. B* 2003;91:1.
[191] J. Ramkumar, B. Maiti. *Separat. Sci. Technol.* 2004;39:449.
[192] J.S. Do, W.B. Chang. *Sens. Actuat. B* 2004;101:97.
[193] C.-J. Yuan, C.-L. Hsu, S.-C. Wang, Ku-Shang Chang. *Electroanalysis.* 2005;17:2239.
[194] L. Doretti, D. Ferrara, P. Gattolin, S. Lora, F. Schiavon, F.M. Veronese. *Talanta* 1998;45:891.
[195] R. Langer, N.A. Peppas. *AIChE J.* 2003;49:2990.
[196] N.A. Peppas, Y. Huang, M. Torres-Lugo, J.H. Ward, J. Zhang. *Annu. Rev. Biomed. Eng.* 2000;2:9.
[197] K.J. Sreenivasan. *J. Appl. Polym. Sci.* 2004;94:651.
[198] F. L. Martien. *Encyclopedia of Polymer Science and Engineering.* Vol. 17. John Wiley: New York, 1986. p 167.
[199] R.Y.M. Huang, C. K. Yeom. *J. Membr. Sci.* 1991;62:59.
[200] C.K. Yeom, K.H. Lee. *J. Appl. Polym. Sci.* 1996;59:1271.
[201] X. Feng, R. Y. M.Huang, *J. Membr. Sci.* 1996;109:165.
[202] A.B. Seabra, L.L. Da Rocha, M.N. Eberlin, M.G. De Oliveira. *J. Pharm. Sci.* 2005;95:994.
[203] Z. Ping, Q.T. Nguyen, A. Essamri, J. Ne'el. *Polym. Adv. Technol.* 1994;5:320.
[204] Z. Ping, Q.T. Nguyen, J. Ne'el. *Macromol. Chem. Phys.* 1994;195:2107.
[205] V. Mano, M.E.S.R.E. Silva, N. Barbani, P. Giusti. *J. Appl. Polym. Sci.* 2004;92:743.

[206] Sanli, E. Orhan, G. Asman. *J. Appl. Polym.* 2006;102:1244.
[207] C.S. Lengsfeld, M.C. Manning, T.W. Randolph. *Curr. Pharm. Biotechnol.* 2002;3.227.
[208] M. Gulati, M. Grover, S. Singh, M. Singh. *Int. J. Pharm.* 1998;165:129.
[209] R.H. Evans, X. Zheng, K.E. Bohannon, B. Wang, M.W. Bruner, D.B.Volkin. *J. Pharm. Sci.* 2000;89:76.
[210] R. Fernandez-Urrusuno, P. Calvo, C. Remunan-Lopez, J.L. Vila-Jato, M.J. Alonso. *Pharm. Res.* 1999;16:1576.
[211] P. Couvreur, G. Barratt, E. Fattal, P. Legrand, C. Vauthier. *Crit. Rev. Ther. Drug Carrier Syst.* 2002;19:99.
[212] C. Witt, T. Kissel. *Eur. J. Pharm. Biopharm.* 2001;51:171.
[213] Y. Yamaguchi, M. Takenaga, A. Kitagawa, Y. Ogawa, Y. Mizushima, R. Igarashi. *J. Controlled Release.* 2002:81:235.
[214] K.F. Pistel, B. Bittner, H. Koll, G. Winter, T. Kissel. *J. Controlled Release.* 1999;59:309.
[215] M. Wittmar, J.S. Ellis, F. Morell, F. Unger, J.C. Schumacher, C.J. Roberts, S.J.B. Tendler, M.C. Davies, T. Kissel. *Bioconjugate Chem.* 2005;16:1390.
[216] E. Chiellini, A. Corti, S. D'Antone, R. Solaro. *Prog. Polym. Sci.* 2003;28:963.
[217] S. Matsumura, N. Tomizawa, A. Toki, K. Nishikawa, K. Toshima. *Macromolecules.* 1999;32:7753.
[218] W.N.E. van Dijk-Wolthuis, S.K.Y. Tsang, J.J.K. den Bosch, W.E. Hennink. *Polymer.* 1997;25:6235.
[219] T. Yamaoka, Y. Tabata, Y. Ikada. *J. Pharm. Sci.* 1995;84:349.
[220] T. Blessing, J.-S. Remy, J.-P. Behr. *J. Am. Chem. Soc.* 1998;120, 8519.
[221] K. Nakamae, T. Nizuka, T. Miyata, M. Furukawa, T. Nishino, K. Kato, T. Inoue, A.S. Hoffman, Y. Kanzaki. J. Biomater. *Sci., Polym. Ed.* 1997;9:43.
[222] T. Jung, W. Kamm, A. Breitenbach, G. Klebe, T. Kissel. *Pharm. Res.* 2002;19:1105.
[223] S.V. Vinogradov, T.K. Bronich, A.V. Kabanov. *Bioconjugate Chem.* 1998;9:805.
[224] K.S. Oh, S.K. Han, Y.W. Choi, J.H. Lee, J.Y. Lee, S.H. Yuk. *Biomaterials.* 2004;25:2393.
[225] Gutowska, Y.H. Bae, H. Jacobs, J. Feijen, S.W. Kim. *Macromolecules.* 1994;27:4167.
[226] P.S. Stayton, T. Shimoboji, C. Dong, A. Chilkoti, G.H. Chen, J.M. Harris, A.S. Hoffman. *Nature.* 1995;378:472.
[227] S.H. Yu, S.H. Cho, S.H. Lee. *Macromolecules.* 1997;30:6856.
[228] F. Illman, T. Tanaka, E. Kokufuta. *Nature.* 1991;349:400.

[229] G. Lorenz, D. Klee, H. Hocker, C. Mittermayer. *J. Appl. Polym. Sci.* 1995;57:391.
[230] N.A. Peppas, P.Bures, W. Leobandung, H.Ichikawa. *Eur. J. Pharmaceutics and Biopharmaceutics.* 2000;50:27.
[231] S. Patachia, S. Varga, R. M. Ion, M. Ranja. Porphyrin encapsulation in nanostructured hydrogels. *J. Optoelectron. Adv. Mater.* 2006 (submitted).
[232] K. Morimoto, A. Nagayasu, S. Fukanoki, K. Morisaka, S.H. Hyon, Y. Ikada. *Pharm. Res.* 1989;6: 344.
[233] T.N. Yudanova, E. Aleshina, E. Obolonkov, I. Dubovnik, L.S. Gal'braikh. *Fibre Chem.* 2004;36:62.
[234] F. Yoshii, Y. Zhanshan, K. Isobe, K. Shinozaki, K. Makuuchi. *Radiat. Phys. Chem.* 1999;55:133.
[235] T.N. Yudanova, E.Yu. Aleshina, M.S. Saenko, L.S. Gal'braikh. *Fibre Chem.* 2003;35:29.
[236] T.N. Yudanova, I.F. Skokova, E. Yu. Aleshina, L. S. Gal'braikh. *Fibre Chem.* 2001;33:20.
[237] K.S. Bohl Masters, S.J. Leibovich, P.Belem, J.L. West, L.A. Poole-Warren, *Wound Repair Regener.* 2002;10:286.
[238] S.L. Bourke, M. Al-Khalili, T. Briggs, B.B. Michniak, J. Kohn, L.A. Poole-Warren . *AAPS Pharm. Sci.* 2003;5:article 33.
[239] M. G. Cascone, M. Laus, D. Ricci. *J. Mater. Sci.: Mater. in Medicine,* 1995;6:71.
[240] E. Jabbari, S. Karbasi. *J. Appl. Polym. Sci.* 2004;91:2862.
[241] H. Kobayashi, M. Kato, T. Taguchi, T. Ikoma, H. Miyashita, S. Sgimmura, K. Tsubota, J. Tanaka. Mater. Sci. Eng. C: *Biomimetic Supramolecular Syst.* 2004;24:729.
[242] Artificial bone connection prosthesis. *United States Patent.* 5314478.
[243] M. Kobayashi, J. Toguchida, M. Oka. *Biomaterials.* 2003;24:639.
[244] M. Kobayashi, J. Toguchida, M. Oka. *Knee.* 2003;10:47.
[245] M. Kobayashi. *Biomed. Metr. Eng.* 2004;14:505
[246] M. Kobayashi, Y.S. Chang, M. Oka. *Biomaterials.* 2005;26:3243.
[247] M.M. Amiji. *Biomaterials.* 1995;16:593.
[248] M.G. Cascone, B. Sim, S. Downes. *Biomaterials.* 1995;16:569.
[249] J.M. Yang, M.C. Wang, Y.G. Hsu, C.H. Chang, S.K. Lo. *J. Membr. Sci.* 1998;138:19.
[250] J.M. Yang, Y.J. Jong, K.Y. Hsu, C.H. Chang. *J. Biomed. Mater. Res.* 1998;39:86.
[251] J.M. Yang, Y.J. Jong, K.Y. Hsu. *J. Biomed. Mater. Res.* 1997;35:175.
[252] T. Chandy, C.P. Sharma. *J. Appl. Polym. Sci.* 1992;44:2145.

[253] V.B. Kushwaha. *J. Appl. Polym. Sci.* 1999;74:3469.
[254] C.K. Yeom, K.H. Lee. J. Membr. Sci. 1996;109:257. J. Ge, Y. Cui, Y. Yan, W. Jiang. *J. Membr. Sci.* 2000;165:75.
[255] C. Peniche, W. Arguelles-Monal, N. Davidenko, R. Sastre, A. Gallardo, J.S. Roman. *Biomaterials.* 1999;20:1869.
[256] P. Gong, L. Zhang, L. Zhuang, J. Lu. *J. Appl. Polym. Sci.* 1998;68:1321.
[257] T. Koyano, N. Koshizaki, H. Umehara, M. Nagura, N. Minoura. *Polymer.* 2000;41:4461.
[258] W.Y. Chuang, T.H. Young, C.H. Yao, W.Y. Chiu. *Biomaterials.* 1999;20:1479.
[259] D.K. Kweon, S.B. Song, Y.Y. Park. *Biomaterials.* 2003;24:1595.
[260] Y. Hu, X. Jiang, Y. Ding, H. Ge, Y. Yuan, C. Yang. *Biomaterials.* 2002;23:3193.
[261] M. Ishihara, K. Nakanishi, K. Ono, M. Sato, M. Kikuchi, Y. Saito, H. Yura, T. Matsui, H. Hattori, M. Uenoyama, A. Kurita. *Biomaterials.* 2002;23:833.
[262] X.Y. Shi, T.W. Tan. *Biomaterials.* 2002;23:4469.
[263] S.J.K. Francis, H.W.T. Matthew. *Biomaterials.* 2000;21:2589.
[264] J.M. Yang, W.Y. Sua, T.L. Leu, M.C. Yang. *J. Membr. Sci.* 2004;236:39.
[265] A.R. Gennaro, G.D. Chase, D.A. Marderosian, S.C. Harvey, D.A. Hussar, T. Medwick, E.G. Rippie, J.B. Schwartz, E.A. Swinyard, G.L. Zink (Eds.). *Remingtonis Pharmaceutical Sciences,* vol. 1151, Mack Printing Company, PA, 1990.
[266] J.M. Rosiak, I.Janik, S. Kadlubovski, M. Kozicki, P. Kujawa, P. Stasica, P. Ulanski. The *International Atomic Energy Agency's* report. © 2002.
[267] I.-K. Kang, J.-S. Moon, H.M. Jeon, W. Meng, Y.I. Kim, Y.J. Hwang, S. Kim. *J. Mater. Sci.: Mater. in Medicine.* 2005;16:533.
[268] F. Lim, A.M. Sun. *Science.* 1980;210:908.
[269] S.J. Sullivan, T. Maki, K.M. Borland, M.D. Mahoney, B.A. Solomon, T.E. Muller, A.P. Monaco, W.L. Chick. *Science.* 1991;252:718.
[270] T.H. Yang, N.K. Yao, R.F. Chang, L.W. Chen. *Biomaterials.* 1996;17:2139.
[271] T.H. Yang, W.Y. Chuang, N.K. Yao, L.W. Chen. *J. Biomed. Mater. Res.* 1998;40:385–391.
[272] W. Paul, C.P. Sharma. *J. Biomed. Sci. Polym. Ed.* 1997;8:755.
[273] T.H. Young, N.K. Yao, R.F. Chang, L.W. Chen. *Biomaterials.* 1996;17:2131.
[274] T.H. Young, N.K. Yao, W.Y. Chuang, L.W. Chen. *J. Biomed. Mater Res.* 1998;40:385.
[275] T.H. Young, W.Y. Chuang, M.Y. Hsieh, L.W. Chen, J.P. Hsu. *Biomaterials.* 2002;23:3495.

[276] K.K. Sirkar, P.V. Shanbhag, A.S. Kovvali. *Ind. Eng. Chem. Res.* 1999;38:3715.
[277] T. Tan, F. Wang, H. Zhang. *J. Molec. Catalysis B: Enzymatic.* 2002;18:325.
[278] V.M. Balcao, A.L. Paiva, F.X. Malcata. *Enzym. Microb. Technol.* 1996;18:392.
[279] F.M. Veronese, C.Mammucari, F. Schiavon, O.Schiavon, S.Lora, F. Secundo, A.Chilin, A. Guiotto. *Il Farmaco.* 56(2001) 541-547.
[280] N.O.V. Sonntag. *J. Am. Oil Chem. Soc.* 1982;59:795.
[281] R. Roxu, U.Y. Iwasaki, T. Yamane. *J. Am. Oil Chem. Soc.* 1997;74:445.
[282] T.W. Tan, F. Wang, T.Q. Liu, *J. Chin. Cereals Oils Assoc.* 2000;2:29.
[283] J. Xu, Y. Wang, Y. Hu, G. Luo, Y. Dai. *J. Membr. Sci.* 2006;281:410.
[284] L. Sarda, P. Desnuelle. *Biochim. Biophys. Acta.* 1958;30:513.
[285] G. Pugazhenthi, A. Kumar. *J. Membr. Sci.* 2004;228:187.
[286] L. Giorno, R. Molinari, M. Natoli, E. Drioli. *J. Membr. Sci.* 1997;125:177.
[287] N.J. Horan, C.X. Huo. *Environ. Technol.* 2004;25:667.
[288] P. Raviyan, J. Tang, B.A. Rasco. *J. Agric. Food Chem.* 2003;51:5462.
[289] Deshpande, S.F. D'Souza, G.B. Nadkarni. J. *Biosci.* 1987;11:137.
[290] J. Rosiak, K. Burczak, J. Olejniczak, W. Pekala. *Polim. Med.* 1987;17:99.
[291] Gupte, S.F. D'Souza. *J. Biochem. Biophys. Methods.* 1999;40:39.
[292] S.K. Jha, S.F. D'Souza. *J. Biochem. Biophys. Methods.* 2005;62:215.
[293] P.A. Ramires, E. Milella. *J. Mater. Sci.: Mater. Medicine.* 2002;13:119.
[294] C.A. Finch. *Polyvinyl Alcohol.* John Wiley and Sons., Bristol, UK, 1973.
[295] J.B. Faisant, A. Ait-Kadi, M. Bousmina, L. Deschenes, *Polymer.* 1998;39:533.
[296] N. Walling, M.R. Kamal, *Adv. Polym. Technol.* 1996;15:269.
[297] S.Y. Lee, S.C. Kim, J. *Appl. Polym. Sci.* 1998;67:2001.
[298] S. De Petris, P. Laurienzo, M.Malinconico, M Pracella, M. Zendron. *J. Appl. Polym. Sci.* 1998;68:637.
[299] L. Xu, L. Zhang, H. *Chen. Desalination.* 2002;148:309.
[300] Y.I. Park, K.H. Lee. *J. Indust. Eng. Chem.* 1999;5:235.
[301] K. Patel, R.S.J. Manley. *Macromolecules.* 1995;28:5793.
[302] J.-T. Yeh, L.-H. Wang, K.-N. Chen, W.-S. Jou. *J. Mater. Sci.* 2001;36:1891.
[303] N.N. Li, E. Drioli, W.S. Winston Ho, G.G. Lipscomb (Eds.). Advanced Membrane Technology. *Ann. N.Y. Acad. Sci.*2003;984.

INDEX

A

acclimatization, 37
acetic acid, 1, 19, 20
acetone, 15, 22
acid, 14, 18, 19, 20, 24, 43, 44, 45, 47, 48, 50, 53, 54, 60, 62, 69
acrylic acid, 8, 14, 15, 21, 47
acrylonitrile, 18, 20
actinic keratosis, 63
activation, 16, 19, 24, 27, 59, 68
activation energy, 16, 19, 24, 27, 59
actuators, 43
additives, 23, 40, 59
adhesion, 14, 57, 61, 71
adhesives, 57
adsorption, 25, 39, 47, 52, 68, 72
aerosols, 54, 76
agent, 3, 5, 12, 14, 17, 20, 45, 59, 63, 69, 72
aggregation, 10
agriculture, 63
air pollution, 72
albumin, 57, 65, 66
alcohol, 1, 8, 9, 10, 12, 13, 14, 15, 18, 19, 21, 25, 26, 27, 34, 35, 40, 44, 45, 47, 48, 52, 54, 60, 61, 65, 70, 75, 77, 80
alcohols, 21, 27
aldehydes, 21, 27
allyl amine, 66
allylamine, 12, 17, 22

alternative, 7, 24, 34, 41
alternatives, 44
amino-groups, 53
ammonia, 66
anhydrase, 72
aniline, 9, 10
animals, 4, 58
annealing, 13, 45
anticancer drug, 63
apoptosis, 55
apparel, 5
aqueous solutions, 5, 8, 9, 14, 15, 17, 18, 61
articular cartilage, 52, 62
ascorbic acid, 50
assessment, 63
asthma, 50
attachment, 31
attention, 43, 47, 48, 53, 68, 72, 75
availability, 25

B

barriers, 76
beams, 1, 58
behavior, 15, 46, 57, 62
bending, 46, 47
benign, 43
benzene, 23, 24, 25, 26
binding, 4, 31, 35, 37, 47, 63
bioaccumulation, 37

Index

bioavailability, 54, 55
biocatalysts, 69
biocompatibility, 14, 40, 52, 63, 76
biodegradability, 5, 37, 63, 64, 75, 76
biological stability, 39
biomass, 13, 37
biomaterials, 51, 52, 55, 63, 70
biomedical applications, 2, 52, 61
biomolecules, 49
biosensors, 43, 48, 50
biosorption, 37
blend films, 29, 47
blends, 8, 14, 15, 17, 19, 21, 23, 25, 28, 35, 37, 45, 52, 63, 70, 73, 77
blood, 49, 51, 52, 54, 65
bloodstream, 55
body fluid, 64
body weight, 4
bonding, 26, 34, 54
bonds, 48
boric acid, 47, 83
by-products, 27

C

Ca^{2+}, 66
calibration, 50
cancer, 34, 50, 55
candidates, 29, 51, 53
capillary, 35
carbon, 23, 26, 49
carbon monoxide, 23
carcinoma, 63
carrier, 34, 53
cartilage, 63
cast, 14, 18, 45
casting, 14, 18, 20, 21, 24, 29, 35
catalysis, 5
catalyst, 20, 24, 66, 67, 68, 76
catalysts, 15, 34, 43, 67
catalytic hydrogenation, 25
catalytic system, 68
cell, 37, 43, 44, 45, 55, 61, 63, 69
cell adhesion, 61
cell culture, 69
cell culture method, 69
cell cycle, 63
cell death, 55
cell growth, 61
cell line, 61
cell organization, 61
cellulose, 17, 39, 48, 57
cellulose diacetate, 48
ceramic, 5, 62
ceramics, 51
channels, 10, 44
chemical interaction, 47
chemical reactions, 70
chemical stability, 52
chitin, 63
chromatography, 38
chronic diseases, 34
circulation, 53
clean energy, 13
clean technology, 7
cleaning, 37
CO_2, 72
coagulation, 41
coatings, 54, 63
collagen, 61, 62
colon, 63
combined effect, 58
compatibility, 29, 49, 53, 64
compensation, 47
compliance, 55
components, 4, 5, 7, 15, 18, 21, 23, 25, 27, 31, 32, 38, 64
composites, 8, 17, 61, 77
composition, 8, 9, 12, 15, 16, 19, 20, 21, 22, 23, 24, 27, 35, 37, 41, 45, 70, 71, 73
compost, 5
compounds, 2, 7, 32, 34, 52, 64, 75
concentration, 2, 3, 16, 17, 18, 22, 26, 29, 30, 32, 37, 38, 46, 47, 48, 49, 50, 52, 71, 72
condensation, 7, 10, 12, 19
conductivity, 9, 43, 44, 45, 48, 49
constant rate, 55
control, 5, 9, 10, 13, 14, 31, 60, 63, 72, 75
conversion, 20, 24
conversion rate, 20

Index

copolymers, 8, 17, 51, 52
copper, 32
cornea, 61
corrosion, 70, 73
cosmetics, 63
coupling, 54
covalent bond, 67
covalent bonding, 67
creatine, 49
creatinine, 49, 63
crystalline, 25
crystallinity, 26, 62, 70
crystallites, 3
culture, 61
curing, 12, 14, 17, 20, 21, 22, 51
cycles, 3, 14, 15, 57, 61
cyclodextrins, 28, 34
cytocompatibility, 69
cytotoxicity, 53, 69

D

death, 4
deformation, 41, 71
degree of crystallinity, 70
dehydrate, 22
dehydration, 7, 8, 10, 12, 13, 14, 15, 17, 18, 19, 20, 21, 22
delivery, 4, 37, 52, 53, 54, 55, 58, 59
demand, 55
demulcent, 5
denaturation, 66
density, 12, 30, 44, 49
deposition, 64
depreciation, 70
derivatives, 18, 48, 59, 68, 70, 76
dermatoses, 63
desorption, 59
detection, 46, 49
detergents, 38
diabetes, 50, 66
dialysis, 38, 49, 51
diamines, 53
diffraction, 46

diffusion, 7, 19, 26, 29, 30, 35, 36, 44, 46, 59, 67, 68
diffusivity, 13
diluent, 5
discrimination, 31
disinfection, 55
dispersion, 10
dissociation, 46
distillation, 7, 13, 21, 24, 25, 27
distribution, 26, 38, 39
DMA analysis, 61
DNA, 38, 53
dopants, 10
doping, 10
dosing, 54
dressing material, 58
dressings, 58, 59
drug delivery, 4, 5, 52, 53, 54, 55, 57
drug delivery systems, 53
drug release, 60
drugs, 51, 55, 57, 58, 75
drying, 49
durability, 20, 40
duration, 55

E

Education, 81
effluent, 25
elasticity, 16, 57
electrical conductivity, 9
electricity, 55
electrocatalyst, 48, 49
electrodes, 48, 50
electrolysis, 9
electrolyte, 36, 43, 44, 52
electromigration, 46
electron, 1, 2, 34, 58
electron density, 34
electron pairs, 34
electrons, 50
electrophoresis, 35
electroporation, 55
electrostatic interactions, 53
elongation, 26, 40

emission, 72
enantiomers, 35
encapsulation, 55, 86
endothelial cells, 61
energy, 7, 19, 21, 24, 34, 43, 55, 72
energy density, 43
England, 77
entrapment, 47, 48, 55, 69
environment, 12, 38, 39, 51, 57, 68
environmental conditions, 37
environmental impact, 5
environmental protection, 34
environmental stimuli, 54
enzymatic activity, 38
enzyme immobilization, 67
enzymes, 34, 49, 50, 53, 67, 68, 75
equilibrium, 46, 63
ester, 48, 54
ethanol, 13, 14, 15, 16, 17, 18, 19, 23, 24, 48, 54
ethyl alcohol, 27
ethylene, 21, 54, 70
ethylene glycol, 21
ethylene oxide, 54
evaporation, 7, 14, 20, 21, 24, 29, 36
exclusion, 36, 50
excretion, 53
exercise, 58
exposure, 9, 12
extracellular matrix, 65
extraction, 2, 5, 31, 32, 38, 54
extraction process, 2, 31
extrusion, 54, 73

F

fabric, 5
fabrication, 40, 49, 50
family, 44
fat, 63
FDA, 5
FEMA, 20
fermentation, 13, 27
fermentation broth, 27
fibers, 2, 20
fibroblasts, 60
fillers, 10, 60
film, 9, 14, 21, 35, 41, 48, 59, 70, 71, 72, 75
film formation, 75
films, 2, 14, 15, 35, 48, 57, 59, 71, 76
filtration, 5, 39, 40, 41, 68
fish, 34
fixation, 61, 62
flexibility, 12, 17, 73
fluid, 38, 66
food, 5, 39, 67, 68, 70
food industry, 39, 67
food safety, 39
formaldehyde, 41, 48, 63
fouling, 39, 40, 41, 50
free volume, 12, 19, 25, 26
freezing, 1, 3, 4, 14, 15, 32, 45, 57
fuel, 23, 43, 45, 72, 73, 76
fuel efficiency, 44

G

gas phase, 37
gases, 75
gasoline, 23, 72, 73
gastrointestinal tract, 4
gel, 1, 3, 5, 13, 30, 39, 46, 48, 52, 54, 57, 60, 67, 76
gelation, 2
generation, 4, 57, 58
glucose, 34, 48, 50, 57, 65
glucose oxidase, 48, 50
glutamate, 49
glycerol, 68
goals, 8
government, 72
grades, 57
graphite, 25, 26
groups, 9, 10, 12, 17, 34, 43, 44, 46, 48, 52, 53, 54, 75
growth, 37, 57, 60
growth factor, 60

H

harmful effects, 69
HDPE, 71, 73
healing, 57, 58, 60
health, 49
health care, 49
heat, 20, 24
hepatocytes, 64, 66
hepatoma, 64
hexane, 54
high density polyethylene, 71, 73
hospitals, 5
host, 34, 48, 61
host tissue, 61
humidity, 70
hybrid, 10, 12, 13, 19, 25, 26, 61, 64
hybridization, 13
hydrocarbons, 73
hydrogels, 1, 2, 4, 14, 15, 30, 32, 34, 46, 51, 52, 54, 55, 56, 57, 60, 61, 63, 64, 75, 80, 86
hydrogen, 26, 34, 50, 54
hydrogen atoms, 34
hydrolysis, 1, 3, 12, 19, 39, 40, 41, 52, 57, 67, 68, 70, 75
hydrophilicity, 9, 14, 17, 40, 48
hydrophobic interactions, 39
hydrophobicity, 14, 44
hydroxide, 46
hydroxyapatite, 61
hydroxyl, 1, 10, 35, 40, 41
hydroxyl groups, 1, 10, 35, 40, 41
hygiene, 5

I

immersion, 39, 57
immobilization, 35, 47, 48, 58, 67, 68, 69
immune system, 64, 65
immunoglobulin, 65
implants, 53
impregnation, 31, 47
in situ, 10, 26
in vitro, 60, 69
in vivo, 53, 65
inclusion, 32, 35, 60
incompatibility, 73
indicators, 47, 48
indomethacin, 57
industrial application, 8, 67
industry, 5, 18, 50, 67
infection, 58
inhibition, 68
inhibitor, 60
injections, 53
insertion, 37
insulin, 57, 65
interaction, 25, 26, 32, 46, 50, 51
interactions, 34, 37, 54, 59
interface, 46, 68, 71
interference, 50
intermolecular interactions, 54
interval, 55
inversion, 41
iodine, 32
ion exchangers, 34
ions, 26, 32, 35, 37, 46, 49, 50
irradiation, 2, 69
isobutylene, 24
isomers, 29, 31, 35

K

K^+, 49
kidney, 49, 64
kidney dialysis, 49
kidney failure, 49
kinetics, 47
knees, 62

L

lamellae, 70, 71
laminar, 71, 73
lamination, 5
landfills, 5
lateral meniscus, 62
lending, 34

lesions, 63
life expectancy, 49
linear molecules, 38
linkage, 48, 67
lipase, 67, 68
lipases, 67, 68
liquid interfaces, 31
liquid phase, 24, 31
liquids, 9, 27, 58
liver, 66
living conditions, 64
low temperatures, 38
LTD, 77
luminescence, 47
lysine, 50

missions, 23
mixing, 14, 18
models, 34
modules, 66
molasses, 27
mole, 31, 39
molecular mass, 56
molecular structure, 48
molecular weight, 3, 8, 38, 40, 52, 56, 59, 60
molecules, 10, 26, 28, 34, 38, 47, 53, 68
monomer, 1, 10, 15, 20
monomers, 75
Moon, 79, 82, 87
morphology, 59, 61, 70, 71
mucosa, 52

M

mammalian cells, 49
management, 43
manufacturing, 27, 50
market, 75
matrix, 1, 10, 14, 15, 25, 32, 37, 48, 49, 52, 54, 55, 58, 61, 66, 67, 68, 69
measurement, 47
mechanical properties, 15, 26, 45, 63
media, 10, 34, 39
medicine, 50, 63, 77
melt, 72
membrane separation processes, 66
membranes, 1, 5, 7, 8, 9, 10, 12, 13, 14, 16, 17, 18, 19, 20, 21, 22, 23, 24, 25, 26, 29, 31, 32, 34, 35, 36, 37, 38, 39, 40, 41, 43, 44, 45, 46, 48, 50, 52, 53, 54, 63, 64, 65, 66, 67, 68, 69, 70, 72, 75, 76
metabolism, 49, 54, 55
metabolites, 54
metals, 51, 67
methacrylic acid, 20
methanol, 1, 22, 24, 43, 45, 48, 72, 73
micrometer, 68
microorganism, 37
microspheres, 1, 53
migration, 30, 61
miniaturization, 43

N

Na^+, 36, 49, 56
NaCl, 21
nanoparticles, 10, 32, 53
natural environment, 5
network, 26, 35, 46, 52
New York, iii
nitric acid, 37
nitric oxide, 60
nitrogen, 12
nontoxicity, 63
nutrients, 64

O

occlusion, 51
oil, 27
oils, 68
olive oil, 68
optical chemical sensors, 47
optical fiber, 20
optical properties, 75
optimization, 12, 17, 49
organ, 64
organic compounds, 7, 22, 46
organic solvent, 5, 7, 12, 17, 21, 33, 67
organic solvents, 5, 12, 17, 21, 33

Index

organism, 64, 69
osmosis, 8
osmotic pressure, 46
ossification, 62
osteoporosis, 50
output, 47
oxidation, 9, 11, 50
oxides, 67
oxygen, 34, 55, 64, 70, 72

P

PAA, 8, 9, 12, 13, 16, 17, 22
pain, 57, 58
paints, 20
palladium, 49, 50
palm oil, 68
PAN, 18, 21
pancreas, 64, 65, 66
particles, 10, 26, 37, 38, 59, 61, 68, 70, 71
passive, 9, 37
PCR, 39
peptides, 51
peritoneal cavity, 64
permeability, 7, 15, 21, 26, 31, 36, 39, 44, 45, 49, 58, 63, 64, 65, 70, 71, 72, 73
permeation, 5, 7, 13, 17, 18, 19, 20, 21, 25, 26, 28, 29, 41, 43, 73
pH, 15, 32, 38, 40, 46, 47, 48, 49, 52, 54, 62
phase inversion, 41, 67
phenol, 8, 9, 47, 48
phenolphthalein, 48
photolithography, 50
photopolymerization, 48
physical properties, 2, 52, 64
plants, 34
plasma, 12, 14, 49, 55
plasma levels, 55
plasmapheresis, 52
platelets, 51
PMMA, 66
poison, 43
polar groups, 8
polarization, 67
pollutants, 31, 32

pollution, 37, 73
poly(vinylpyrrolidone), 45
polyacrylamide, 19, 69
polymer, 1, 2, 3, 4, 7, 8, 9, 10, 12, 14, 15, 17, 18, 22, 25, 29, 31, 34, 35, 37, 38, 39, 40, 41, 43, 44, 46, 48, 49, 50, 51, 52, 53, 54, 57, 58, 61, 63, 69, 72, 75, 76, 77
polymer blends, 45, 69
polymer chains, 10
polymer materials, 14, 54
polymer matrix, 10, 35, 38, 48
polymer mixing, 15
polymer networks, 15, 51, 63
polymer structure, 7
polymeric blends, 52
polymeric materials, 52
polymeric membranes, 21, 26, 34
polymerization, 1, 9, 10, 14, 15, 57, 69
polymerization mechanism, 9
polymerization process, 69
polymers, 9, 14, 15, 35, 39, 43, 44, 45, 50, 51, 52, 53, 54, 57, 67, 73, 75
polyolefins, 70
polypropylene, 71
polyurethane, 54
polyvinyl alcohol, 48, 49, 62, 68
poor, 9, 20, 43, 52, 53, 54, 63, 71, 73
population, 34
porosity, 36, 39, 40, 62
porphyrins, 34, 55, 56, 80
power, 43
precipitation, 37, 66
preference, 57
pressure, 7, 12, 21, 38, 39, 46
production, 27, 50, 75
prognosis, 62
proliferation, 61, 69
propylene, 54
prosthesis, 62, 86
proteins, 38, 48, 49, 51, 53, 64, 66
proteolytic enzyme, 58, 59
protons, 43, 46
PTFE, 68
pure water, 35
purification, 34, 35, 39, 72, 75

Index

PVA, 1, ii, 1, 2, 3, 4, 5, 8, 9, 10, 12, 13, 14, 15, 16, 17, 18, 19, 20, 21, 22, 23, 24, 25, 26, 27, 28, 29, 30, 31, 32, 33, 34, 35, 36, 37, 39, 40, 41, 44, 45, 47, 48, 49, 50, 52, 53, 54, 55, 56, 57, 58, 59, 60, 61, 62, 63, 65, 66, 67, 68, 69, 70, 72, 73, 75, 76, 80, 83
PVA films, 59, 70
PVAc, 1, 3
PVP, 45, 52

Q

quartz, 46

R

radiation, 1, 61, 69
radical polymerization, 69
range, 4, 5, 14, 15, 25, 37, 40, 43, 46, 49, 50, 66
reactant, 66
reactants, 66, 67
reaction medium, 67
reaction rate, 69
reactivity, 70
reagents, 69, 75
recognition, 4, 34, 54, 61
reconstruction, 52
recovery, 25, 35, 38, 68
rectum, 63
reduction, 12, 19, 47, 55
regenerated cellulose, 39
regeneration, 37
regulations, 39
rejection, 21
repair, 62
replacement, 62, 72
reproduction, 4
resection, 62
residuals, 69
residues, 2, 48
resistance, 17, 35, 52, 64, 68, 73, 75
response time, 47, 48
retention, 34, 37, 38, 39, 41

RNA, 38
rods, 1
Romania, 80
room temperature, 29, 45

S

safety, 4, 58, 73
sales, 75
salt, 10, 21, 46, 57, 59
salts, 21, 26, 27, 38
sample, 36, 46, 71
search, 10
sedimentation, 14
seeding, 61
segregation, 66
selectivity, 7, 8, 9, 12, 14, 17, 19, 21, 22, 25, 26, 31, 34, 72
self-control, 64
semiconductor, 50
sensitivity, 46, 47, 49, 75
sensors, 43, 46, 47, 49, 76
separation, 1, 5, 7, 8, 9, 12, 13, 14, 16, 17, 18, 19, 20, 21, 22, 23, 24, 25, 26, 27, 28, 29, 31, 35, 36, 38, 52, 63, 66, 67, 72, 75, 76, 81
serum, 39, 41
serum albumin, 39, 41
shape, 1, 35, 38, 41, 52
shape-memory, 52
shear, 38
shock, 62
Si_3N_4, 49
side effects, 53, 55, 63
silicon, 46, 47
similarity, 61
SiO_2, 49
skin, 52, 55, 57, 58
skin diseases, 55
society, 76
sodium, 21, 35, 36, 59, 72
soil, 5
sol-gel, 13, 26
solubility, 5, 13, 18, 25, 52, 57, 68
solvents, 15, 35, 40, 73
sorption, 19, 29, 30, 34, 37, 45, 56, 59

Index

species, 5, 35, 37, 38, 49, 50
spin, 39, 49
stability, 1, 9, 12, 17, 21, 40, 44, 47, 48, 49, 50, 52, 53, 67, 72
stock, 57
storage, 49
strength, 3, 16, 17, 20, 21, 40, 48
strong interaction, 10
structural characteristics, 52
styrene, 35
substrates, 47, 48
sucrose, 27
suffering, 51
sugar, 27
sulfuric acid, 1
Sun, 78, 80, 87
supported liquid membrane, 31
surface modification, 10, 13
surface properties, 14
surface region, 10
survival, 65, 66
susceptibility, 9
swelling, 1, 3, 4, 10, 13, 15, 17, 19, 27, 29, 35, 44, 46, 53, 54, 57, 59, 61, 63, 69
swelling process, 46
synergistic effect, 26
synthesis, 2, 5, 9, 10, 13, 25, 39, 51, 66, 67, 68, 75
synthetic polymers, 52, 57, 69
systems, 4, 5, 34, 38, 51, 53, 55, 59, 63, 64, 66, 70

T

tacticity, 52, 75
technology, 5, 8, 31, 39, 54, 55, 75
temperature, 3, 8, 9, 12, 15, 16, 17, 18, 19, 22, 23, 24, 27, 41, 45, 46, 52, 53, 68, 70, 72
tensile strength, 16, 57, 69
TEOS, 12, 13, 19
tetraethoxysilane, 13
textiles, 5
therapy, 34, 55
thermal resistance, 40, 52
thermal stability, 39, 40, 44
thermal treatment, 14
thyroid, 49
time, 2, 3, 35, 36, 46, 47, 58, 59, 60, 62, 68
tissue, 4, 54, 55, 61, 62, 63, 64, 66
toluene, 23, 24, 71, 72
toxic effect, 69
toxic metals, 37
toxicity, 3, 4, 37, 52, 76
trade, 21
transistor, 49
transition, 54
transitions, 54
transmission, 63, 70
transplantation, 65
transport, 8, 19, 26, 46, 55, 63
transportation, 43
trifluoroethyl methacrylate, 20
trypsin, 60
tumor, 63
tumor cells, 63

U

UK, 88
uniform, 55
unit cost, 50
United States, 86
urea, 49
urethane, 53, 54
uric acid, 50, 63
UV, 14, 60
UV irradiation, 14

V

vacuum, 7, 29
validity, 54
values, 19, 32, 45, 47, 59
vanadium, 49
vanadium pentoxide, 49
vapor, 69, 70, 73, 76
variable, 37
variation, 47
vehicles, 56, 60

vein, 61
viscosity, 59, 71
vitamin B1, 63
vitamin B12, 63

washing procedures, 69
waste water, 31, 32, 80
wastewater, 5, 32, 37, 39, 55, 75
wastewater treatment, 5, 32, 39
water absorption, 44

water diffusion, 19
water evaporation, 29
water sorption, 59
water vapor, 59
weakness, 18, 47
wear, 62, 63
weight ratio, 15
wetting, 49

yield, 68